Ernest Jacob

Notes on the Ventilation and Warming of Houses, Churches,

Schools and other Buildings

Ernest Jacob

Notes on the Ventilation and Warming of Houses, Churches, Schools and other Buildings

ISBN/EAN: 9783337014056

Printed in Europe, USA, Canada, Australia, Japan

Cover: Foto ©berggeist007 / pixelio.de

More available books at **www.hansebooks.com**

MANUALS OF HEALTH.

NOTES ON THE

VENTILATION AND WARMING

OF

HOUSES, CHURCHES, SCHOOLS,

AND OTHER BUILDINGS.

BY THE LATE

ERNEST H. JACOB, M.A., M.D. (Oxon.),

PROFESSOR OF PATHOLOGY, YORKSHIRE COLLEGE, LEEDS.

PUBLISHED UNDER THE DIRECTION OF THE GENERAL
LITERATURE COMMITTEE.

LONDON:

SOCIETY FOR PROMOTING CHRISTIAN KNOWLEDGE,

NORTHUMBERLAND AVENUE, W.C.; 43, QUEEN VICTORIA STREET, E.C.

BRIGHTON: 135, NORTH STREET.

NEW YORK: E. & J. B. YOUNG & CO.

1894.

[While this little book was going through the press its gifted author was called away from the world. His life, full of promise, was brought to an end on March 1st, 1894.]

INTRODUCTION.

THE present work is an attempt to put into popular form information respecting matters of health which are very little understood by the general public. These subjects are considered in technical works on engineering, and in bulky treatises on Hygiene, written for the use of Medical Officers of Health and Sanitary Engineers; but these books are not generally accessible to the ordinary reader. Certain excellent chapters on the subject of ventilation may be found in some recent Manuals on the subject of Health, but mixed with much extraneous matter, such as directions for diet and exercise. The present notes are the result of considerable practical experience in examining and dealing with insanitary buildings, and the remedies recommended have all been practically tried and proved successful.

It is hoped that the book will be found useful by the general householder no less than by those who are interested in the sanitation of schools, churches, and similar buildings.

The author wishes to express his thanks to Prof. Goodman, of the Yorkshire College, for assistance in the revision of the engineering formulæ in the chapters on calculation methods, and to his various friends of the architectural profession, and others who have allowed him to use their drawings to illustrate the work.

Acknowledgments are also due to the Clarendon Press; Messrs. Crosby Lockwood, & Co.; Messrs. Churchill; Kegan Paul & Co.; and Mr. Jackson, for the loan of engravings, or permission to use illustrations of which they possess the copyright.

E. H. J.

Leeds,
  January 1894.

CONTENTS.

ILLUSTRATIONS.

NOTES

VENTILATION OF BUILDINGS.

—•◦•—

CHAPTER I.

THE NECESSITY OF VENTILATION.

ENTILATION is a term applied to the method by which a due supply of fresh air is maintained in buildings and other confined places, such as mines and ships.

The word was invented about two hundred years ago by Dr. Desaguliers, a well-known scientific man of the time, who devoted a great deal of time, money, and inventive skill towards improving the sanitary condition of buildings, as well as of ships, the condition of which, even at that time, was considered very bad. In the course of his work Dr. Desaguliers invented a "fanning wheel" (Fig. 1), or what we now call a "fan" or "air propeller," which was worked by manual power, and the man who worked the fan was called the "ventilator." The word has since been divorced from its original meaning, and used to denote a hole in a building through which air may (or may not) pass, according to circumstances.

Every one knows that a supply of air is necessary for all living things, a few of the lowest vegetable organisms excepted. To cover the head with a cloth for a short time produces a sense of oppression, and the removal of the veil is followed by a feeling of relief. This sense of oppression is increased by the heat which is given off from the breath, and there is

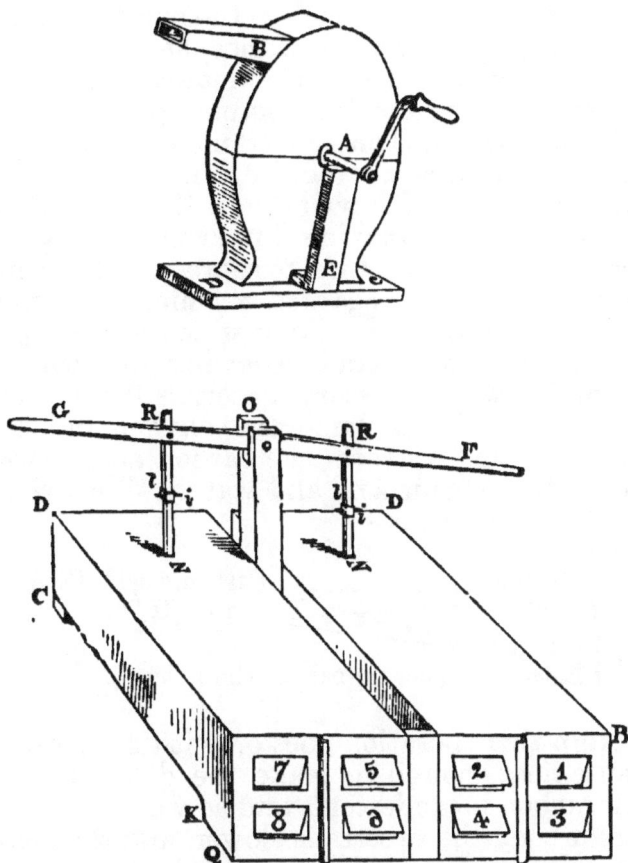

Fig. 1.—Desaguliers' "Blowing Engine," and Hales' Ventilating Bellows (1734).—*Tomlinson.*

considerable confusion in the minds of most people as to the relative effects of the fouling and the mere heating of the air by respiration. People commonly say of a crowded building "it is too hot," whereas most of the discomfort comes from the impurity of the atmosphere. There is no sense of oppression felt in a well-ventilated Turkish bath, although the temperature may be very high.

Air, or rather oxygen, which is the effective agent in the mixture of gases we call atmospheric air, is absolutely necessary for life. The maintenance of our bodily heat, energy, and the complex processes necessary for "life," depends on a constant supply of oxygen. The other substances we require, viz. carbon, hydrogen, nitrogen, &c., which we take in the form of food, can be supplied at varying intervals. Oxygen we must have continually. Life, to represent the complex by the simple, may be regarded chemically as similar to the burning of a lamp, a lamp which has the power of making its own oil and renewing its own wick, when supplied with necessary materials for the manufacture. In the lamp flame the oxygen of the air combines with and converts the carbon and hydrogen of the oil into carbonic acid and water. We write the process chemically thus—

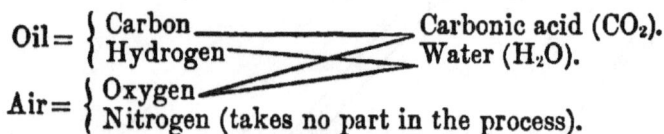

$$
\begin{aligned}
&\text{Oil} = \begin{cases} \text{Carbon} \\ \text{Hydrogen} \end{cases} \qquad \begin{array}{l} \text{Carbonic acid } (CO_2). \\ \text{Water } (H_2O). \end{array} \\
&\text{Air} = \begin{cases} \text{Oxygen} \\ \text{Nitrogen (takes no part in the process).} \end{cases}
\end{aligned}
$$

In a lamp this operation goes on with such rapidity as to cause heat enough to make the disengaged particles of carbon incandescent, and so we get light. In the human body the process is slower, and the temperature remains at about 98° Fahr., but the essence of the process is the same, and the products are also mainly carbonic acid and water.

A glance at the analysis of the air before and after it has passed through the lungs will show how great is the change produced by respiration, and how very foul the expired air becomes. Pure air consists of a mixture of about 21 parts of oxygen, 79 of nitrogen in 100, some watery vapour, a trace of ammonia, and carbonic acid 4 parts in 10,000. Expired air contains not only about 5 per cent. less oxygen, but carbonic acid in the enormous proportion of 470 parts in 10,000, and is moreover heated to the temperature of the body, about 98°, and is saturated with moisture.

In addition to these changes, expired air contains an organic impurity of perceptible odour, which we recognize as a "stuffy smell," and there are exhalations more or less odorous from the skin.

There is considerable difference of opinion among physiologists as to the precise impurity which produces the deleterious effect which is felt in crowded rooms. Some experiments made in Paris a few years ago, seemed to show conclusively that the principal agent was the organic impurity referred to above. Recently, however, these experiments have been repeated, both in Germany and England, with negative results ; and it seems probable that, as was originally held, the excess of carbonic acid must be credited with the injurious effects noted. This is upheld by the fact that the effects produced in a room crowded with people are very similar to those perceived when much gas is burnt in an unventilated room. No doubt, however, the unpleasant sensations are intensified by the effect on the delicate sense of smell of the disagreeable exhalations referred to above. However this may be, an estimation of the carbonic acid present in a given sample of air is, speaking generally, a fair test of its purity.

The fumes of burnt gas have just been referred to, and in fact this source of impurity is an extremely important one. Besides carbonic acid and water, which are produced by the combustion of coal-gas, sulphurous acid is given off, and if the gas flame be cooled to any extent, as occurs in cooking-stoves, a very pungent gas called acetylene is formed. In addition, recent investigations seem to show the presence, in minute quantities, of carbon monoxide, which is one of the most poisonous gases known, and, as an ingredient of coal-gas, is the active agent in the many deaths which have been caused by the accidental escape of gas in houses.

Air may be tested for carbonic acid roughly by the following method.

Six stoppered bottles are taken, containing respectively 450, 350, 300, 250, 200, and 100 cubic centimetres. These are filled with the air of the room which has to be tested by means of a small hand-ball syringe. A glass tube or pipette, holding exactly 15 cubic centimetres, is then filled with clear transparent lime-water and emptied into the smallest bottle, which is then shaken. If the fluid becomes turbid, the amount of carbonic acid will be at least 16 parts in 10,000. If no turbidity appears, repeat the operation with the next largest bottle. Turbidity will here indicate 12 parts. In similar fashion, turbidity in the 250 c.c. bottle indicates 10 parts CO_2 ; in the 300 c.c. bottle 8 parts ; in the 350 c.c. bottle 7 parts, and in the 450 c.c. bottle less than 6 parts. To judge of the turbidity, mark a piece of paper with a lead-pencil, and gum it on to the bottle with the mark inside. If there be turbidity the mark will be invisible.

The method usually adopted by chemists is to take a large jar, holding about a gallon—the large "spice jars" used by confectioners answer very well. This is filled with the air to be examined by means of bellows. 100 c.c. of lime-water are put in, and the bottle shaken and allowed to stand. The alkalinity of the lime-water is then ascertained by measuring the amount of oxalic acid which is required to neutralize it before and after it has been shaken up with the air.

The oxalic acid solution is made by dissolving 5·64 grms. of crystallized oxalic acid in a litre of water, and is of such strength that 1 c.c. will exactly correspond with 1 c.c. of carbon dioxide. An example will show the manner of working.

50 c.c. of the lime-water required 34 c.c. oxalic acid solution
for neutralization.

,,　　,,　　,,　　(after
shaking with air) required　32 c.c.　　,,　　　,,

Difference　2　　= c.c. of CO_2 in the
lime-water used.

As the 30 c.c. was only half that used in the test,
the number must be doubled, *i. e.* 4 c.c.

The capacity of the bottle was 4100 c.c.

∴. 4000 c.c. (*i. e.* 4100 − 100 c.c. taken up by the lime-
water) (4 litres) of air contain 4 c.c. of carbon dioxide
= 1 per cent., or 10 parts in 10,000. Solution of phenol
phthalein is generally used to indicate the point of
neutralization.

For further accuracy it is advisable to make a cor-
rection for temperature, the standard weight being
given at zero Centigrade. For this, ·002 per cent.
should be added for every degree above 32° Fahr.
Corrections for pressure are less necessary. For further details the reader is referred to special works on
chemical analysis.

The following figures give an average view of the
insanitary state of most of our public buildings. The
figures are given in parts per 10,000,—4 being the
standard for the open air, and 6 the best obtainable
ventilation for a building.

Analyses made in Leeds by Professor Thorpe.

	CO_2.
Sitting-room—near floor　...　...　...	7·33
,,　　half-way up　...　...　...	9·
,,　　near ceiling　...　...　...	14·65
Grand Theatre, Leeds—pit　...　...　...	15·01
,,　　,,　　upper circle　...　...	14·29
,,　　,,　　balcony ...　...　...	14·16
Philosophical Hall—after lecture...　...　...	13·38

Analyses in Nottingham by Professor Clowes.

Grounds of University College　...　...　...	4·3
Chemical Laboratory of University College　...	6·9

	CO_2
Masonic Hall during a dance	31·
Circus	32·6
Committee-room with 15 persons and 27 gas-jets ...	41·8

Dr. Angus Smith gives the following from Manchester.

Theatre Royal—pit	27·34
,, gallery	13·58
School-room	9·7
Mills (400 people) (1)	28·6
,, ,, (2)	29·6
,, ,, (3)	30·

London.

Chancery Court	19·3
Olympic Theatre	10·
Standard Theatre	32·

We may remark, with reference to the above figures, on the very excellent ventilation of the Chemical Theatre at the Nottingham University College, considering that at the time it was filled with students, each using a Bunsen burner. The fact that the upper gallery of a theatre may be purer than the pit may be explained that the pit, generally very crowded, is often largely covered over, and the ceiling low, while, the fumes being attracted to the central chandelier, the air from the lower parts penetrates the upper galleries to a very slight extent.

The effects of breathing respired air have been tested by various observers, the person making them being placed in a closed chamber, or inspiring and expiring into an india-rubber bag. The susceptibility of various individuals varies greatly. The following is a brief account of the researches of Dr. Angus Smith, who had a special leaden chamber constructed for the purpose. Considering that conditions very nearly similar to those produced by Dr. Angus Smith are daily present in hundreds of assembly-rooms, churches, and other buildings, it is well that we should note his results. Speaking of bad air and its effects, Dr. Smith says—

"Here I am describing feelings, and to some persons they may simply be fancies, but I shall describe them nevertheless, as I believe man has learnt nearly all he knows of ventilation by attention to these feelings, while chemical analysis is attempting to struggle after him, and is continually finding itself behind him in the race.

"The first trial of the chamber was made by simply sitting down 1 hr. 40 min. This produced about 1 per cent. of CO_2 (carbonic acid gas). No difference was to a certainty perceptible for 25 minutes. Then when the air was drawn from the top by means of an umbrella, it seemed like a soft wind, and had to some extent a pleasant feeling, but was entirely devoid of a faculty of cheering.

"After an hour the unpleasant smell of organic matter, such as is so well known in a crowded school, was perceptible on movement. It was decidedly perceived, after remaining an hour, that the air was very soft when made to move. This arose from the moisture, and shows us at once that a soft air may be a very impure one. Soft air with a good deal of vapour is very soothing; it calms the mind and the body, and the burning of a candle and a fire. After staying in the chamber for 100 minutes, the air had an unpleasant flavour or smell, and I came out. Three persons entered at once, and pronounced it very bad. It seemed to me, however, that we are frequently exposed to air equally bad, though I have never found any in daily life so much deprived of its oxygen (20 per cent). I was very glad of the escape from this impure air, the gladness not arising from any previous discomfort. There was unusual delight in the mere act of breathing, which feeling continued for four hours.

"The second stay in the chamber was continued for 160 minutes. After 140 minutes it was observed that very long inspirations became frequent, and more agreeable than usual. The air about that time had a very decided feeling of closeness. Immediately on opening the door two or three persons entered, and again perceived how uncomfortable it was."

Experiments were then made on the combustion of candles, and it was found that in one instance four miners' candles went out after five hours' burning, and again that eight candles, one paraffin lamp, and one spirit-lamp were all out in 150 minutes. The room was then entered, the persons carrying candles and a spirit-lamp. The lights were soon extinguished, and it was found impossible to kindle them with matches. Nevertheless, they could breathe, though every one was glad to go out. No very correct description of their feelings could be given.

Dr. Smith adds : "All these experiments tend to diminish our faith in our feelings as guides under certain conditions. The senses are quite unable to measure degrees of closeness." The importance of this observation can hardly be over-estimated. "A young lady was anxious to be in the chamber when the candles went out. She was very fond of pure air, but was not much struck with the impurity of the air in the chamber, though the candles were threatening to go out, so that there could not have been quite 19 per cent. of oxygen, with 2·1 per cent. of CO_2. She stood for five minutes quite well, but suddenly became white, and could not come out without help."

It is impossible to read the accounts of these experiments without being reminded of what we may at any time feel and see in a crowded church at an evening service, or many theatres, though the latter are, as a rule, better ventilated than churches.

Not many years ago a colossal suffocation experiment was made in a large town in the north of England. This, though less disastrous than the well-known catastrophe of the "Black Hole of Calcutta," was on a much larger scale, and quite sufficiently unpleasant to those who took part in it.

A great politician was expected, to make an important speech. As there was no room of sufficient dimensions available in the town, a large courtyard, surrounded with buildings, was temporarily roofed over, some space being left under the eaves for ventilation. Long before the appointed time several thousand people assembled, and in due course the meeting began ; but before the speaker had got well into his subject, there arose from the vast multitude a cry for air, numbers of people were fainting, and every one felt oppressed and well-nigh stifled. It was only after some active persons had climbed on the roof and forcibly torn off the boards for a space about twenty feet square, that the business of the meeting could be resumed. This occurrence so exactly illustrates the ignorance prevailing on the subject of ventilation, that it deserves wider publicity than it obtained at the time. We would point out that the question of the provision of fresh air was not forgotten, the fault lay in the builder's ignorance of the laws governing movements of air. And yet this method of so-called ventilation is in use (and useless) in numbers of buildings at the present day.

The general effects of exposure to air contaminated by the products of respiration or the burning of gas, may be summed up as the following—

Headache, sometimes felt at the time, and of long duration, but occasionally not coming on till some hours afterwards, perhaps the next morning. It then takes the "migraine" or sick headache type, with its accompanying symptoms of vertigo, intolerance of light and sound, and occasionally vomiting. Besides headache, there is generally sleepiness, lassitude, inability to fix the attention, and loss of appetite. It is not uncommon to hear clergymen complain of feeling "Mondayish," as they term it. On Monday morning they are accustomed to suffer from exhaustion, headache, and a feeling of malaise. This is attributed to the hard work of Sunday, but the real reason is to be found in the intolerable atmosphere breathed on

Sunday night in a crowded church, and there are many cases to show that an improvement in the atmosphere of the church entirely removes the weekly headache, with no curtailment of work.

The symptoms and sensations described above are frequently experienced by those who only occasionally are exposed to foul air, as in theatres, churches, &c. The effects on the health of those who habitually live and work in crowded and unventilated buildings are unhappily only too well known. Sir John Simon wrote in 1863, in one of two series of masterly essays on public health which characterized his administration as the sanitary adviser of the Privy Council : " In proportion as the people of a district are attracted to any collective *indoor* occupation, in such proportions (*ceteris paribus*) the district death-rate from lung disease will increase." In that year (1863) the deaths from consumption in the country districts being taken as 100, the deaths in Manchester counted 263, and in Leeds 218. The greatest mortality took place among printers and tailors, classes who work largely by night, requiring a strong light, which necessitates the burning of much gas. On the other hand, contemporary statistics showed that the miners of Northumberland and Durham, where the pits were freely ventilated, formed an important exception to this rule. The necessity of keeping mines free from explosive gas has forced the managers to employ powerful ventilation apparatus, and a recent writer (*Reports of the Laboratory of College of Physicians, Edinburgh,* 1891), after a careful investigation of the health of a large mining district in Scotland, finds that with the improved atmospheric conditions the miners' liability to pulmonary consumption has disappeared. Workshops are still far from perfect, though vastly improved by the more general use of ventilating fans. The white faces of the working girls in too many of our great towns still tell a sad story, although an army of government inspectors are commissioned to report on any insanitary conditions. But "quis custodiet custodes?" The inspectors are conscientious, well-educated gentlemen, but not skilled in sanitary knowledge.

Next, as to the amount of air required. A person standing in the open air, on a calm day, is exposed to about 32,000 cubic feet of air passing by him per hour. It is out of the question that he should be supplied with this amount in a closed space, but careful experiments on barrack-rooms by the late Dr. Parkes, Professor of Hygiene at the Military Medical School at Netley, have shown that the best room ventilation which can be reasonably available will supply 3000 feet per head per hour. The air will then remain absolutely without sensible odour, and the carbonic acid impurity will not exceed 6 parts in 10,000. This is ideal ventilation for rooms inhabited the whole day, as hospitals, &c., and this amount is generally provided in good hospitals for cases of infectious diseases.

In the case of rooms inhabited for shorter periods, churches, assembly-rooms, and the like, a much smaller amount of air will suffice, but the smallest amount which could be called even moderately good ventilation would be not less than 500 cubic feet per head per hour. This means that an ordinary sitting-room 16 × 16 × 12 ft., containing about 3000 cubic feet, would require for continuous use the air to be changed once an hour for one person, three times an hour for three persons, but if used for merely an hour or two at a time, it would be fairly wholesome with eight persons, if the air be changed three times an hour, and there be no gas burnt. It would be, however, a difficult matter in so small a room to effect this without causing draughts.

In rooms occupied for a short time only the amount of air required per head varies according to the size of the room. The following table from Parkes shows the amount required during the first hour under these circumstances, the full amount of 3000 cubic feet being required for the second hour.

Amount of air required to dilute to standard of 6 parts of CO_2 in 10,000.

Cubic space per head.	Air required for 1st hour.	Subsequent hours.
100	2900	
200	2800	
300	2700	
400	2600	
500	2500	3000
600	2400	
700	2300	
800	2200	
900	2100	
1000	2000	

The following table represents the generally-accepted standard of good ventilation.

	Cubic feet per head per hour.
Hospitals (general)	3000
,,　　 (infectious)	5000
Theatres	2000
Assembly-rooms	2000
Prisons	1760
Workshops (ordinary)	2000
,,　　 (unhealthy) ...	3500
Barracks (day)	2000
,,　 (night)	2000
Schools (adult)	2000
,,　 (infant)	1000
Stables	6000

When the proportion of carbonic acid in a room, as the result of respiration, is increased from the usual proportion of 4 in 10,000 to about 8, a faint musty odour can generally be detected by any one entering the room from the outside air. As a rule, it may be said that if the atmosphere of a room is quite free from unpleasant odour to a person entering from a fresh atmosphere outside, there is very little fault to be found with the ventilation of the room as far as the effective change of air is concerned. This musty smell, however, is

frequently masked by stronger odours, flowers, per-
fumes, &c., and then an air analysis must be referred
to. The amount of air required for comfort depends
largely on the temperature of the external air, as well
as the dimensions of the room. With regard to the
latter, it is generally allowed that very lofty rooms are
not desirable; they are difficult to warm, and there is a
kind of reservoir of cold air at the top. Ordinary
living-rooms should not be more than 14 feet high, and
a little lower is preferable. With regard to the difference of temperature, Dr. Billings remarks : "It will be
found that when the outside air is below the freezing-
point, and the room has the usual proportion of
external wall and window space, the amount of air-
supply per hour shall be about 1½ times the cubic
contents of the room, otherwise either the room will
not be kept warm, or the fresh air will have to be
introduced at a much higher temperature than is
desirable for health or comfort." This is more applicable to the conditions found in the Northern States of
America, where not only is the winter temperature
very low, but the average temperature preferred in a
living-room is about 10° higher than that considered
sufficient in England. American houses are generally
kept at about 70°. Dr. Billings further remarks : "The
higher the external temperature, the more air is required for comfort. There are some days in summer
when sufficient air to secure comfort can hardly be
obtained even in the open air, and the feeling of
having insufficient air is often felt in a crowd, though
in the open air."

The amount of crowding allowable in a room is a
rather important question. Practically, as there is
very little efficient ventilation in most buildings, it is
the practice to allow a comparatively large area per
individual. If, however, there is a regulated supply of
air, a point is reached at which the due supply of air
would be required to move at too great a velocity for
comfort, say more than five feet per second, so that we

must really restrict the crowding even of well-ventilated rooms to that point at which sufficient air can be given at a velocity not exceeding five feet per second. One occasionally sees a court of law so crowded that an adequate supply of air would render it difficult for the counsel to retain their wigs.

Some explanation must be given of the reason why so large an amount of air is required, considering the small amount we actually inspire. We may illustrate by the analogy of a water-supply.

Engineers allow a minimum of twenty gallons a head in estimating the amount of water to be supplied to a town. The water is taken from a tap into a basin, and having been used, is discharged into a drain. The fouled water does not mix with the common supply. Now imagine a household of ten persons, with the twenty gallons per head in one central cistern, from which all water is to be taken and to which it must be returned. Something of this kind obtains in many village communities in India, but in civilized countries would be considered an inexpressibly foul arrangement. If there were no sinks or waste pipes, nothing short of a rapid stream of many thousand gallons a head through the house would meet the difficulty. A precisely similar condition of things exists with reference to our air-supply in a confined space. The air we breathe is taken from a common stock, and breathed back with all its gaseous impurities, laden with moisture and heat, into the common stock, which rapidly becomes foul, unless constantly removed. The prodigious amount of air we require is not for the purpose of supplying us with oxygen, but in order so to dilute the poisonous substances produced by respiration, that they become innocuous and free from odour.

We have mentioned some of the evils incidental to an insufficient supply of air. It seems hardly necessary to state the converse of this as an incentive to the obtaining of proper sanitary conditions. A few may be given as instances.

In the early days of sanitation, the operatives in a mill in which a fan had been erected for ventilating purposes applied to the proprietors for increased wages; on the plea that they ate so much more food! But they also did much more work. The same remark as to work has been frequently observed by school-masters. An experienced master lately observed to the writer : "In the old school" (where the rooms were close and unventilated) "the boys used to get tired about four o'clock, but here, in our new buildings, they work on well to the end." In this school a carefully-designed system of ventilation changed the air in the rooms three times an hour.

In healthy places of worship there are no sleepers among the congregation, and the minister wakes on Monday morning none the worse for his Sunday's work.

With the ventilation of mines and workshops, the mortality from consumption has steadily decreased.

By the ventilation of the stables of the French cavalry the mortality among the horses was reduced from 197 per 1000 to 20. "A horse seldom takes 'cold' from exposure to cold, but frequently is made ill from being too warm," says Major Fisher (*Through the Stable and Saddle-room*). "It is the inside, not the outside air that gives them coughs, sore throats, congestion of the lungs, and sundry other ills to which horseflesh is heir."

The same applies to cows, and all animals which are kept in confinement, as the directors of the Zoological Gardens have found to their cost.

Some interesting observations have been made by Carnelly and Haldane in a recent examination of school buildings in Dundee ventilated by mechanical means and otherwise. They found that on examining the air of the school-rooms for micro-organisms, a very much smaller proportion were found in the schools mechanically ventilated. More than this, when the ventilation in these schools was intermitted for a time, it was found that on resuming the working of the fan, the air still remained comparatively free from micro-organisms. The reasonable explanation of this is, that the ventilated rooms contained little which could act as a nidus or appropriate soil for these organisms to inhabit. (*Vide* Appendix.)

The micro-organisms here referred to are not necessarily germs of disease. Many are perfectly harmless. Many produce what we call putrefaction when growing in organic matter, and may be described as Nature's scavengers. Still, where the conditions for breeding micro-organisms exist, those producing disease may be found with the rest, and there is no doubt that pure air is comparatively free from these forms of life. As a matter of fact, the air of a well-ventilated sewer contains fewer micro-organisms than many a school-room or workshop.

X

CHAPTER II.

THE PRESENT INSANITARY STATE OF BUILDINGS.

WE have seen that the living body breathes, that is, takes in oxygen and expels carbonic acid, some fifteen to twenty times a minute during its whole life. It is obvious that buildings which are to hold living bodies must breathe also, although the precise mechanism by which the change of air is effected may be different. When we come to examine by this standard existing buildings, we find things by no means satisfactory. Builders know that wooden floors must be ventilated, or the "dry rot" fungus, the *Merulius lacrymans*, will feed on their substance and destroy them; but the idea that breathing human beings in a confined space require any special arrangements for supplying them with air at a convenient temperature has seldom taken a practical form—churches, houses, schools are being built every day in which the subject is entirely neglected. Concert-rooms and theatres are slightly better, inasmuch as it is usual to use "Sunlights," or other ventilating gas-burners for lighting purposes ; but the buildings are seldom fitted with any properly-devised machinery for the purpose. Exception may be made in the case of certain recently-erected theatres, in which mechanical power is used, and school-boards are beginning to adopt similar measures to render healthy schools in large towns, but in most cases architects are content to introduce an occasional air-brick, or a patent contrivance called a "ventilator," which is speedily stopped up, and people stifle in silence, apparently possessed with the idea that it is impossible to introduce fresh air into an inhabited room without its being felt as a draught. The worst offenders against the laws of health are those responsible for the building of churches and other places of worship. The reason of this is not far to seek.

A church is built on a conventional plan, fixed in mediæval times, when churches were less crowded, services shorter, and above all, at a time when there was no lighting by gas. As every point about the structure of a church has been settled by a fixed authority, it is very difficult to introduce changes, and the plans for a new church pass the authorities all the more easily if they are of the conventional type. The usual form of church building, copying as it does more or less closely the beautiful architecture of the thirteenth or fourteenth centuries, is by no means an ideal form from a sanitary point of view, whatever it may be from other considerations. It is generally built in the form of a nave and side aisles, lighted by clerestory windows. This gives, including the chancel, four ceilings of three different heights, making it most difficult to extract the air at the level of the roof. The clerestory windows chill the warm air as it rises, and send it down in the form of a cold douche on the heads of the congregation. The roof is lofty and dark, necessitating a large amount of light, and as a rule about twice as much gas is burned for lighting purposes as is necessary. The pillars obstruct the view of the preacher and the altar from at least ten per cent. of the congregation. The building is too frequently used but once a week, and is therefore hurriedly warmed at the end of the week, an operation often very imperfectly performed ; and last but not least a degree of economy is generally exercised both in the erection and management which is fatal to the obtaining of perfect sanitary conditions.

Bad as churches generally are found to be when examined from a sanitary point of view, Nonconformist chapels are generally worse, on account of the frequency of galleries and the consequent crowding. Worst of all are probably the numerous mission-rooms which, through the energy of the clergy, are found in so large numbers in the poorer districts of our large towns. These are frequently extemporized out of a couple of cottages. No architect is consulted on the subject, the alterations are made by some local builder, and sanitary conditions are absolutely unthought of. The strictest economy is observed, especially in the heating apparatus, which is generally a small stove, and every Sunday a large class of more or less unwashed children is succeeded by a crowd of totally unwashed adults, till the atmosphere of the rooms can only be described as sickening.

It is proposed in the following pages to lay down the general principles by which buildings are rendered healthy, with some practical hints on the treatment of such rooms as the above, on which, owing to their temporary nature, it is not desirable to expend much money. Iron churches are as bad as mission-rooms, and have besides the disadvantage of being intolerably hot in summer. Sunday-schools vary, but are usually very bad, except when they consist of small class-rooms heated by an open fire.

CHAPTER III.

CONDITIONS NECESSARY FOR GOOD VENTILATION.

WHAT now do we mean by speaking of a room as "ventilated"? Real ventilation is so uncommon that there is no general popular consensus on the subject. The architect usually thinks this object has been attained if some of the windows can be opened. Some think that the presence of "ventilators," especially if they have long names, and are secured by "Her Majesty's letters patent," ensures the required end. We may as well attempt to supply a house with water by making a trap-door in the roof to admit the rain.

The answer given by sanitarians to the question of "What is a ventilated room, suitable for human beings to inhabit in comfort and health?" is very definite as regards general principles.

Three conditions must be fulfilled—

1. The building must have its walls warmed to the temperature at which it is required the air should be kept, otherwise a person near the wall will feel cold.

2. A supply of air, in quantity depending on the number of people (speaking generally for rooms not continuously occupied, about 1000 feet per head per hour), must be caused to pass uniformly through the room, at a velocity not exceeding five feet per second.

3. This air must have its temperature so modified by heating or cooling apparatus that, while it gives rise to neither cold nor hot draughts, the temperature of the room remains constant. Other refinements may be added, such as arrangements for filtering and moistening the air, which will greatly add to the comfort of the occupants.

Although this is easily stated, the problem of carrying out these points, especially the third, in rooms of varying size, under circumstances of varying degrees of crowding and great differences of temperature, is by no means easy.

It is very necessary to observe the great importance in this scheme of the heating apparatus. In the climate of England the cooling of the air will be a luxury only occasionally required, and reserved for theatres and places of public resort, where expense is not so carefully considered. Warming, however, is necessary during more than half the year, during nearly three-quarters of the year in the northern parts of the British Isles ; and the fact that heating apparatus is not, as a rule, designed to assist ventilation, either as regards its size or its position, is the principal reason why so few buildings are either healthy or comfortable. The general public, led by advertisements, invests largely in ventilators ; these are found to cause draughts and are immediately closed, the fact being that it is impossible to introduce sufficient fresh air into a crowded room in cold weather, unless the air be heated.

When any attempt is made to improve the air-supply to any completed building, it will be invariably found that the principal difficulty lies in connection with the heating arrangements. As a matter of fact it is impossible to carry out a perfect system of ventilation for a house or other building, unless the heating and ventilation arrangements are designed by the same person ; in fact, the heating must be completely subordinate to the ventilation. Unless this is so it is impossible to obtain good results, and the reason why the ventilation of most buildings is a failure, consists entirely of the fact of the heating apparatus not having been designed for this purpose. The amount of radiating surface is calculated on a basis which may fulfil the first condition referred to above, *i. e.* the warming of the walls of the room, but cannot possibly fulfil the last and most important, *i. e.* the warming of the incoming air.

A glance at the ordinary method of procedure will show how very unlikely it is that satisfactory sanitary results can be obtained.

An architect is instructed to prepare plans for a church, school, or similar building. These are presented to the building committee, and are criticized entirely from the point of view of the architectural appearance and the general convenience of the arrangements. Occasionally some too active member of the committee ventures to ask the architect if he has made any arrangements for ventilation. He always receives the reply, " that the subject has received the most careful attention, and that when the building is finished it will be found perfect in that respect." The inquisitive member of the committee subsides, suppressed. The plans are then, in the case of an ecclesiastical building, submitted to the Bishop, and through him to the diocesan architect. The latter considers carefully the structural arrangements, in order that the building may be well built and durable ; he may make inquiry as to the probable nature and situation of the heating apparatus, with a view to considering any risk from fire which may be possible, but the sanitary condition of the building, as a receptacle for a number of living, breathing, air-consuming human creatures, is entirely neglected. The plans are now passed, and the building is commenced. In the course of its erection the architect selects certain makers of heating apparatus to tender for the warming of the building, and the relative merits of hot air and hot water are discussed. If economy is a principal object, a cheap hot-air apparatus is ordered, and this is the most unwholesome machine which can possibly be put in a church. If wiser counsels prevail, hot-water pipes are fixed ; but as these are generally placed beneath the floor in channels, where they become speedily covered by a coating of non-conducting dust, a much larger amount of piping is required than if they were placed above-ground, and there is generally an unpleasant smell from the burnt dust. Finally the church is finished and consecrated, and mutual congratulations are exchanged. Unhappily, it soon becomes apparent that, except in the summer, when the windows can be kept open, the building is, especially at an evening service, much too hot and very "stuffy." There is an offensive smell from the gas, in spite of the fact that it is lowered during the sermon, to allow the audience to slumber more peacefully under the influence of the increasing quantity of carbonic acid. Eventually, the congregation is divided between those who prefer rheumatism and bronchitis with open windows, and those who prefer asphyxia with the windows shut.

The next stage is the calling in of some maker of patent ventilators, who disfigures the church inside and out with metal tubes and other unsightly contrivances. But although in mild weather these make a slight improvement, on the first cold day such a cataract of draught is felt, that the new ventilators are soon all closed, and once closed they remain so. The congregation mournfully resign themselves to the conviction that ventilation is an impossible thing. Under the circumstances it must be feared they are not far wrong.

This is no fancy sketch. Cases of this kind may be found in plenty.

Although something can be done to improve a building in which these mistakes have been made, as will be seen on reference to the examples given further on, it is impossible to render such a church or other building sanitarily satisfactory without a complete reconstruction of the heating, and probably of the lighting arrangements. The contractor for the warming apparatus undertook only to maintain the building at a certain temperature in the absence of ventilation, and he carries out his contract strictly. He has nothing to do with ventilation unless he receives due instructions on that point—and a "ventilating engineer" can do nothing without the heating engineer.

As an illustration of the great difference between mere heating and what is necessary for ventilation purposes, the case of a large school recently built may

be quoted. A ventilation scheme was carefully devised
by a special committee of the managers who were
familiar with the subject, and a tender obtained from
an experienced heating engineer. This gentleman said,
as he handed in his contract, "If you wish me to *heat*
the building in the usual style, I will do so for rather
less than half the sum named in the tender."

Good ventilation cannot be carried out without the
expenditure of money, and is unfortunately rather
costly, the principal item of expense being the increased amount of heat required. The cheapest way
of heating a room is to fill it with people, close every
opening, and let them "stew in their own juice," and
this is what is frequently done, with the effect on
health mentioned in the first chapter.

It is found that when people are crowded together
in a large town, it is necessary to have elaborate
engineering arrangements to supply them with water,
and communities cheerfully and confidently expend
millions in bringing pure water from distant reservoirs,
the cost being distributed by means of a rate. It is
equally true that when people are crowded together
in a room, they require a supply of pure air artificially
supplied to them, and the cost of this should be as
cheerfully paid as that of water or gas.

We have attempted in the preceding pages to
point out why the method generally adopted in the
erection of a church or school fails to give satisfactory
results. Let us give, on the other side, a sketch of
a different method, drawn from recent experience.

The building to be erected was a large school, to be
used for various technical purposes, containing laboratories, lecture-rooms, museums, &c. As soon as the plans
and elevations had been approved of, a small committee was appointed to consider with the architect
the general principles of the ventilation and heating.
They decided on steam as the means of heating, and a
fan, actuated by a gas engine, as the means of ventilation. The architect was now commissioned to consult

an engineer of special experience in this subject, and by him, in consultation with the architect, a complete scheme was drawn up, the position of every flue was determined, the amount of heating surface for every room calculated on the basis of the amount of ventilation necessary for that room, and every detail arranged before the building was commenced. Moreover, during the erection of the building a representative of the engineer consulted was frequently present to ensure that the projected constructive arrangements were properly carried out. The architect here wisely shared his responsibility with the sanitary expert. This is the method which gives the best, and indeed the only good results. Through the rapid increase of knowledge on sanitary subjects, the architectural profession has burdens laid on it heavier than it can bear, and it is only by the co-operation of architectural and sanitary experts that we can hope to erect buildings on a level, not only with the artistic taste, but also with the sanitary knowledge of the day. Though we boast of our advance on other nations in sanitary matters, the ventilation of public buildings is much more carefully considered on the Continent than in England. Mr. Bacon (Robin's *Technical School and College Building*) remarks that in this respect the Belgian architects do not neglect the matter even if their clients do so, and it is not long since M. Bacckelmans of Antwerp actually refused to carry out the erection of the town hospital—a building of considerable importance—because the Hospital Commission would not appoint an engineer to consider the plans with him, with regard to the heating and ventilation, before the foundations were laid.

Is it too much to ask that in the case of ecclesiastical buildings, in the matter of which there is so much skilled supervision of plans, that the Bishop should number among his advisers a sanitary as well as an architectural expert, and that no plans for a church should be passed unless there were good reason to

o

believe that the building when erected would be a healthy one in use? On expressing this opinion recently to two members of the Episcopal Bench, the author received the reply that it would be impossible for the Bishop to reject the plans unless he was prepared to instruct the builders of the church by what method it was to be ventilated. The Bishop might as well refrain from recommending an oak door because he was unfamiliar with the details of joiners' work, or a bell, because he did not know the precise composition of the alloy used. If there be a demand for professional knowledge on this point, there will be no lack of a supply.

It is in the first erection that the necessary provision can be easily and economically made. It has been well said, "It is easy to make a building breathe if it is caught young." The provision of proper channels and openings is then a simple matter, and the increase of cost is mainly in consequence of the larger amount of heating surface necessary.

CHAPTER IV.

METHODS OF WARMING.

CONSIDERING the important part taken by the heating apparatus in any scheme of ventilation, it will be well to discuss a little more fully the methods in use for warming buildings.

Heat, as will be seen on reference to the general principles laid down in the first chapter, is required for two purposes: (1) For warming the walls of the building, and (2) for heating the incoming air. We must not forget, however, the very large amount of heat which is contributed by the bodies of the living occupants of the building, an amount so as to render it necessary for the air which is supplied for ventilation purposes to be frequently cooler than the air in the building by a few degrees, or the temperature inside rises unpleasantly if the building be crowded. Inasmuch as the same apparatus may be used for both purposes (viz. heating the building and the freshly-supplied air), the general question of ventilating resolves itself into two factors, viz. the heating and the air propulsion. It is true that for a few months in the year heating is not required, but assembly-rooms, schools, and the like are used so much more frequently in the winter, for entertainments, &c., that heat will be required on the majority of occasions on which the building is in use.

Heat, as usually available, is of two kinds, Radiant and Convected. Radiant heat is that which streams like light in straight lines from heated objects. You can hide the light of a candle from your face by a screen. In the same way you can screen the radiant heat of a fire. Like light also, radiant heat decreases with the square of the distance, so that a source of radiant heat, like a fire, is of very little use in warming a large space. We have here to use convected heat, which is communicated from the lower parts of a room to the upper by means of the air. In a small room with a fire the floor and walls are warmed by the radiant heat, and they in turn warm by convection the air in the room. The pleasantness of a fire-warmed room consists in the fact that the walls and objects in the room are warm, while the air is comparatively cool, and the air, being warmed by surfaces not highly heated, does not become dry. When hot-water pipes are used for heating purposes, the amount of radiant heat is very small, but the air surrounding the pipes becomes warm, the colder and heavier air around then forces it upwards, till by the circulation of currents of air the whole space becomes warm. There is more radiation from steam pipes, and more still from a highly-heated stove, but the higher the temperature of the heating surfaces the more dry does the air become, while the organic matter contained in the air becomes charred, and gives out

a disagreeable smell. It is the fashion to call hot-water
and steam coils "radiators." This is a misnomer, as
the amount of radiant heat they give out is small.
They are really air heaters, but there is no other
convenient term to apply to them.

The usual methods of heating a building are—
1. *Open fires.* 2. *Stoves,* or "hot-air apparatus." 3.
Hot-water pipes : high and low pressure. 4. *Steam pipes.*

Bearing in mind the double function of the heating
apparatus, it is obvious that open fires are of very little
use in connection with large rooms. They do not
warm the air required for ventilation, although when
fitted with a hot-air arrangement, as will be explained,
they can be of some use in that matter. On the other
hand, they are the best possible agents for heating a
small room, where the air is not required to be changed
more than once an hour. Much depends, however, on
the kind of open fireplace in use. The open fire is a com-
promise between heating and ventilation. It is possible
to greatly increase the heating power by increasing
the convecting surfaces and slowing the combustion
and draught, as in the "Nautilus" grate and the
"Front Hob" grate of the Teale Fireplace Co. On
the other hand, by using none but radiant heat and
quickening the draught, as in too many of our modern
grates, in which the sides of the grate are of iron, the
ashpit is open, and the chimney is at the back, we may
reduce the heating effect to a minimum, and obtain
only a large flow of air through the chimney.

The kind of grate therefore which is adapted for a
certain room must depend on the amount of ventilation
required, remembering always that if a rapid combustion arrangement is used, there must be some
auxiliary source of heat, whereby the entering air may
be heated. The tendency in some quarters seems to
be of late to reduce rapidity of consumption by check-
ing the draught. This may be carried too far, and the
age of draughts be succeeded by a period of stuffiness.

When fitted with a chamber which communicates

with the external air, the open fireplace enlarges its
sphere of usefulness considerably. The "hot-air," or
"ventilating grate" as it is called, was invented by
Gauger in France some two hundred years ago
(Fig. 2). The best known form of it in England is the

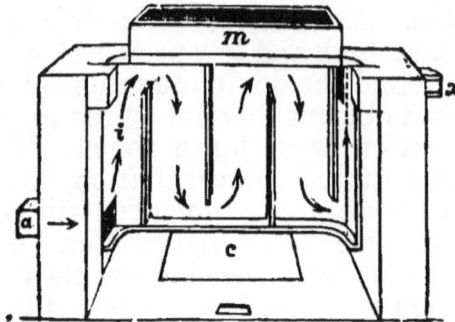

Fig. 2.—The first Ventilating Grate by Gauger.

model designed by Galton, but many good patterns are
made by Boyd, Shorland, the Teale Fireplace Co. (Fig. 3),
and others. Here air is taken from the outside, warmed
by the waste heat in a chamber behind the grate, and
poured warm into the room. The great increase of
heating power obtained in this way, without in any
way increasing the consumption of coal or reducing
the value of the grate as a radiator, shows how wasteful is the ordinary fireplace. The "ventilating grate"
is used in all military hospitals and many other places,
especially schools and small hospital wards, but is not
nearly so much known as it deserves.

Speaking generally, an open fireplace should have a
form differing very little from that laid down by
Rumford nearly a hundred years ago, and lately
advocated anew by Mr. T. P. Teale of Leeds. The
sides and back should be of firebrick, the back inclined
forwards, so that the smoke leaves the fire in front
rather than at the back. The rapidity of combustion
is regulated by the size of the chimney, and the

supply of air to the ashpit.   With our present knowledge it seems impossible to lay down fixed rules

Fig. 3.—Hot-air Grate, with concealed exit.—*Teale Fireplace Co.*
  A, elevation showing perforated overmantel panel to admit
  warmed air. *B,* section showing warming-chamber behind
  grate.

as to what this rapidity should be.   It must depend
on the amount of ventilation required.   Fig. 3*a* indi-

cates the direction of the air currents in a room heated by a fire, and shows that over the fireplace is a convenient position at which to introduce fresh air.

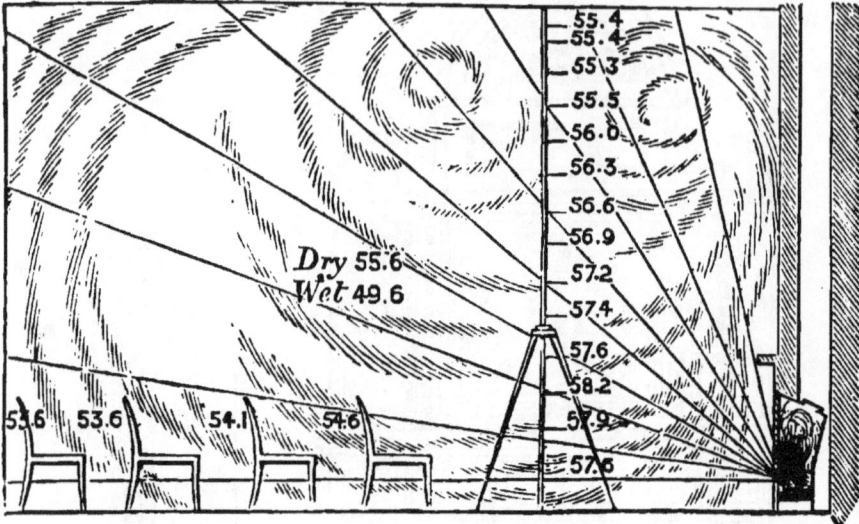

Fig. 3*a*.—Sketch of Experiment made by Mr. Campbell in 1857, showing the movement of currents of air in a room with an open fireplace.—GALTON'S *Health Dwellings.*

II. *Stoves.* This term includes every form of closed apparatus in which the air passes over surfaces directly heated by a fire, from the old-fashioned "cockle" to the latest form of American "furnace." They are extremely economical, if well constructed, and where economy is desired at the expense of comfort or health a stove will generally be adopted.

A stove may be used to heat in four different ways—

1. Standing in the building, surrounded freely by air. In this position it gives off a great deal of heat, by which the air in the room and eventually the walls are warmed. It is thus well adapted for heating a warehouse or storeroom, where no special ventilation is required.

2. If the stove stand against an opening in an external wall, or be furnished with a tube conducting
air from the outside, its heat is expended partly on
the air in the room, and partly on the incoming fresh
air. In this way it acts partly as a simple heater,
partly as a ventilator.

3. The stove may stand in a chamber communicating
with the external air and with the building. This is
the general way in which it is arranged for heating
large buildings, and efficient ventilation *may* be
obtained with this manner of heating, if the following
conditions are strictly adhered to.

(*a*) The stove must be very large,—much larger than
the makers' catalogues recommend, so as to allow of its
giving adequate heat without the surfaces being over-
heated. This will involve greater first cost, but great
economy in working, as it will burn much less fuel and
require very little attention.

(*b*) A suitable exhaust must be arranged for the
building, or otherwise a due amount of air will not be
admitted. If the force which draws in the air be the
excess of temperature in the apparatus chamber over
that of the exterior air, no considerable amount will
be introduced unless the temperature in the chamber
be very high, and this means that the heated air will
be dry, and tend to cause great discomfort.

4. Another way of fixing a furnace is to place it
underground, but to arrange channels so that the
air is drawn from one part of the building into the
heating chamber and discharged, heated, into the
building again. This practically comes to the same
thing as when the stove is placed in the building
itself. There is no possibility of ventilation, since to
make an apparatus of this kind work well there should
be no external openings, and it is in every respect a
most unwholesome arrangement for churches and
buildings occupied by people. Yet it is only too common. Methods 3 and 4 are, however, by some heating engineers combined, and there is no objection to

the latter plan being adopted merely for heating the
building before it is occupied, if fresh air only be
admitted to the heating chamber as soon as the building
is sufficiently warmed. Still, taking them at their best,
stoves, by whatever fancy name they may be called,
are very unsatisfactory things, for two reasons : (1st)
the surfaces are too highly heated for comfort, and
the air is dried unpleasantly; (2nd) it is impossible to
make a cast-iron furnace gas-tight. Experiments by
Putnam (*Open Fireplaces in all Ages*) have shown that in
the best form of American iron hot-air apparatus there
is a constant outward pressure of one to eight milli-
metres of water under ordinary circumstances, and
that there is never an inward pressure, except when
the draught up the smoke flue is too powerful to be
left with safety. The same author gives a most amusing
account of a furnace, guaranteed so confidently by its
maker to be absolutely gas-tight, that he was allowed
to test the matter by experiment. The simple procedure of filling the apparatus with water showed the
presence of innumerable apertures. "To the complete
astonishment of the proprietor, and of the careful
workmen standing around, the water which was poured
in poured out again through nearly every one of the
score of careful joints, until the furnace seemed to dis-
solve and float away in its own tears" (*Op. cit.*, p. 109).

Practically, it is found that very few buildings heated
by "hot air," *i. e.* a stove apparatus, are free at times
from a most oppressive atmosphere, and were it
possible to take a census of the number of persons who
yearly faint in public buildings, it would be found that
in nearly all of these the heating was by some form of
hot air. The only place where these are indispensable
is in a Turkish bath, where as a rule the ventilation is
tolerably good, and a high temperature is required,
very difficult to obtain by any other method. Mr.
Putnam's experiments show that soapstone is impervious to gas at ordinary pressures, and it is much to
be regretted that furnaces of that material which

are procurable in America should not be obtainable in England. The gas which escapes more or less from every cast-iron furnace, consists largely of the very poisonous carbon monoxide referred to above.

III. Hot-water pipes heated by a boiler are the most generally available agents for heating churches and large buildings. There are great advantages in a good hot-water system. The heat is steady, it is impossible to over-heat the pipes, and the warmth is available exactly where it is wanted. It is under perfect control, and the circulation goes on till the pipes are cold. The method is readily adaptable to ventilation schemes. There are two methods in use, viz. the High and Low-Pressure systems. In the former, large pipes (three and four inches in diameter) are used, and the boiler is generally of the type of that used for steam. The temperature of the pipes seldom exceeds 150° Fahr. In the high-pressure system the pipes are very small, and calculated to resist a high pressure. The boiler is a coil of iron pipe placed in a furnace. The circulating water is very small in amount, and the heat is at its maximum very soon after the fire is lighted. The pipes become very hot, considerably above the temperature of boiling water. This renders a very small amount of piping sufficient, and the cost is accordingly much lower than that of a low-pressure system. Certain precautions are necessary to guard against fire, but if the apparatus is properly erected, and the furnace sufficiently large, it acts fairly satisfactorily. There is also a modification of this called the system of "limited high pressure," which has some advantages. All systems of hot-water heating are readily adaptable for ventilation purposes, as the pipes may be either collected in a large heating chamber, or distributed in a number of small ones, and are available for simple heating where no ventilation is required.

IV. Until recently steam was very little used for warming houses or large institutions, its use being mostly confined to workshops, where the waste steam from the engines was available. Methods, however, have been so improved by the practice of American engineers, that the system works with very little trouble, and there is now no reason why·this plan should not be widely used for general heating purposes. It is especially convenient when power is required for some purpose connected with the building, as, for instance, to actuate a ventilating fan, to produce electric light, or work laundry machinery, for the same boiler then suffices as for the engine. Steam coils can be put in any position, and the temperature being high, they need not be large. They are easily shut off by a simple tap, but they are difficult to control, the coil, if it work silently, being either hot or cold. Methods for regulating the heat, however, are in use, and will be referred to later. The cost is slightly under that of a low-pressure hot-water system, but above that of a high-pressure arrangement. A steam boiler, however, requires rather more skilled stoking than a hot-water boiler, and careful inspection after it has been in use some years.

Considering generally the various methods of heating large buildings, it may be said that if a steam boiler be required for any purpose, it will probably be the most economical and convenient to use steam also for heating purposes. Otherwise a low-pressure hot-water system, if erected with due care, leaves little to be desired. The arrangement, however, of the heating surfaces is of great importance. The methods of arrangement are usually classed as (1) *direct*, where the heating surface is placed in the rooms to be heated, and (2) *indirect*, where the radiators are placed so that air passes over them into the rooms. It is obvious that the first method excludes the possibility of ventilation according to the principles laid down above, while the second, or indirect method, lends itself to most methods of good ventilation. It is possible, however, in some cases to combine the two with advantage, as in most large establishments there are some rooms which require very little ventilation, and here the placing of the heating surface inside the rooms has little to be set against it. This is termed the " Direct-indirect " method.

Of late· there has been great improvement in the forms of radiators. Unsightly coils of pipe, which require hiding by ornamental gratings, are giving way to decorative forms of radiator, which can be daily cleansed. Cased coils are receptacles for dust, which becomes baked, and gives off an offensive smell. This is especially the case when pipes are put in channels under the floor, covered with gratings, an arrangement very common in churches. It is objectionable from many points of view. (1st) The pipes give off but comparatively little heat, and therefore an increased amount of piping is required. (2nd) They are soon covered with a layer of dust, which, besides giving off offensive odours, forms a non-conducting coating to the pipe, hindering the due discharge of heat.

It is very important that the boiler should be of large size, so that the maximum heat may be obtained with moderate firing, at conveniently long intervals. The tendency is for the contractor to cut down the expense to the lowest point by tendering for the smallest size boiler which can keep the pipes hot. This means constant labour in firing, and great waste in fuel.

The same applies to the estimation of heating surface. It is far better to have a large surface of low temperature, if space can be given, rather than to have to keep the water at the highest temperature possible. The heating surface, as described above, should be placed in chambers through which the air passes on its way into the rooms. These chambers should be accessible for the purpose of cleaning, as well as of repair to the pipes. In America, "indirect" heaters are usually placed between the floors, horizontally, and are accessible from the ceiling of the room below. In England shallow coils placed in recesses in the wall of the room to be heated are more common, and the casing in front may ópen with hinges like an ordinary cupboard. In the case of a room requiring a good deal of heat and very little ventilation, *e. g.* a room with north aspect, occupied by only a small number of people, there is no objection to an open coil, furnished with an air-supply from behind or underneath. Many radiators are now made with this arrangement.

It is most important that there should be complete control of the heat. This may be obtained in two ways. (1) By arrangement of the coils or radiators in

Fig. 4.—American Nason Radiator, arranged in sections.

three or more sections (*vide* Fig. 4), such a number being used as affords the necessary amount of heat. This is the best way when a large amount of heating surface is concentrated in a special chamber. (2) Where a coil fixed against an inlet in the wall is used for heating and ventilation, the apparatus shown in Figs. 5 and 6 will be found useful. The coil is enclosed in a box furnished with a diaphragm running vertically throughout. By rotating the valve (D) at the upper part the stream of air is allowed to pass through the outer

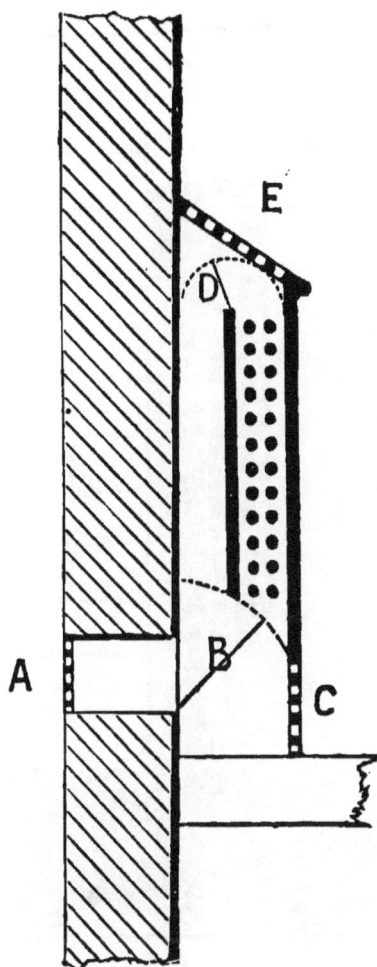

Fig. 5.—Author's Mixing-valve Box for wall radiators.

channel unheated, or over the coil so as to become warm, or by placing the valve in an intermediate position the air may be admitted of any required temperature. By closing the inlet valve (B) at the base, the coil may be used entirely for heating purposes, the air circulating through the front opening (C),

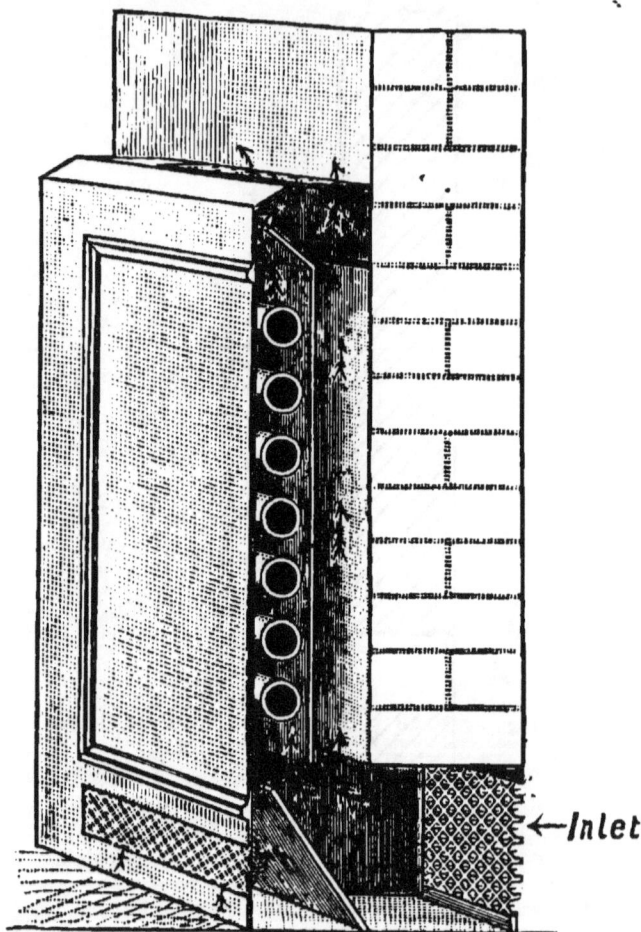

Fig. 6.—Author's Mixing-valve Coil-box for wall radiators.

ventilation being set up the moment the valve is opened. This "mixing-valve" arrangement has been thoroughly tested, and is found to work well under appropriate conditions ; the method is equally available for hot-water and steam apparatus. Without a control arrangement of some kind, steam-heated buildings are generally, in this country, intolerably hot, except in the coldest weather.

To sum up : when a room is filled with people, heat is only required for ventilation purposes (" indirect radiation "). When empty, "direct radiation," or an uncased radiator, is best. When partially filled, a radiator partly open and partly ventilating ("direct indirect ") is suitable. The estimation of the relative amount of "direct" and "indirect" heating surface is a matter requiring much judgment and experience.

The following table, from the *Records of the Smithsonian Institute*, shows the comparative advantages of the various systems described above.

Form of Apparatus.	Percentage of Thermal effect.	Remarks.
1. Open fireplaces ...	10—12	Carry off fresh air, and so bring in fresh but cold air, causing draughts.
2. Ventilating fireplaces	33—35	Carry off foul air and introduce warmed fresh air.
3. Metal stoves without air-supply ...	87—90	Produce no change of air— unhealthy.
4. Stoves with air-supply	68—93	Admit as a rule insufficient air and too highly heated. Bad if cast iron, better if wrought iron.
5. Hot-water pipes ...	65—75	} Easily adaptable for ventilation purposes.
6. Steam pipes ...	58—63	}

CHAPTER V.

METHODS OF LIGHTING.

IT was stated in the first chapter that the air in inhabited rooms was greatly fouled by the burning of gas. Unless the lighting be by electricity—by far the best method—the manner by which a building is lighted is of great importance from a sanitary point of view. If a small number of good oil lamps be used, there will be very little in the way of excessive heat or objectionable smell, and it will only be necessary to allow fresh air to each lamp equal to that required for about three persons.

Candles are so seldom used on a large scale that we may omit their consideration. On the other hand, lighting by gas, which is the most common method, involves considerable modification of the ventilation scheme. According to the way the burners are arranged, they may enormously increase the amount of fresh air required, and accordingly the cost ; or the lighting may be made to assist most usefully the necessary change of air.

Considering the extensive use of gas for lighting purposes, and the frequent exhibitions of gas apparatus held by municipal authorities, there are few subjects on which the public are more utterly ignorant. The ordinary gas-fitter knows nothing about the question of lighting. The knowledge of how to burn gas to the greatest advantage is in the hands of a few, mostly managers of gas-works and makers of gas-burners, who are not as a rule consulted by the public. The competition of the electric light has recently greatly improved the methods of gas lighting, as well as drawn attention to the comparative cost of gas and electricity; but it is impossible to compare these without stating the particular kind of gas lighting with which the comparison is made.

D

Lighting by gas may be done on three different scales of expenditure, with very different results.

1. The ordinary method employed in lighting domestic dwellings and churches is by a number of small burners of common type, *e. g.* Bray's "Regulator," No. 3 or 4. The gas is burnt (very wastefully) at the pressure supplied by the company, which is always far above that at which the burners give their best light. This is the usual method employed by gas-fitters, and on this basis a foot of gas burnt per hour gives a light of about 1½ to 2 candles.

2. The gas may be burned in "checked" burners, of a better type, *e. g.* Bray's "Special" or "Special Slit Union," the pressure being regulated by a governor properly adjusted. A better arrangement still, where only few lights are used, as in a small house, is to use the excellent automatic governor burners made by Sugg, Peebles, and others. The light then averages nearly 3½ candles per foot, a little above the Parliamentary standard.

3. Regenerative burners may be used, in which a great increase of lighting power is obtained by heating the air-supply by means of the waste heat of the burner. These are known by various names, "Wenham," "Cromarty," &c., and give a light value of 9 candles per foot of gas. There is also a class of what may be called "Semi-regenerative," like the "Deimal," and there is a fourth class of burner, the "Incandescent," or "Welsbach," from which a light of extraordinary beauty, with an economy even greater than with the Regenerative, can be obtained; but owing to the fragility of the incandescent material or "mantle," they must at present be considered as on their trial.

Since, therefore, to obtain the light of 100 candles we may burn either 50, 28, or 11 feet of gas, according to the character of the burner employed, it is obvious that the amount of ventilation rendered necessary by the lighting arrangements must vary greatly. To take a well-marked instance. A church was recently inspected by the author as to which there was great complaint on the score of ventilation. The church seated two hundred persons. Now on a liberal computation 100 feet of gas per hour burnt on the standard scale, *i. e.* 525 candles, would be sufficient to properly light the building. Allowing 1000 feet of air per head and 150,000 feet for the gas, the total ventilation requirement would be 380,000 feet. Examination, however, showed that in reality 1800 feet of gas per hour

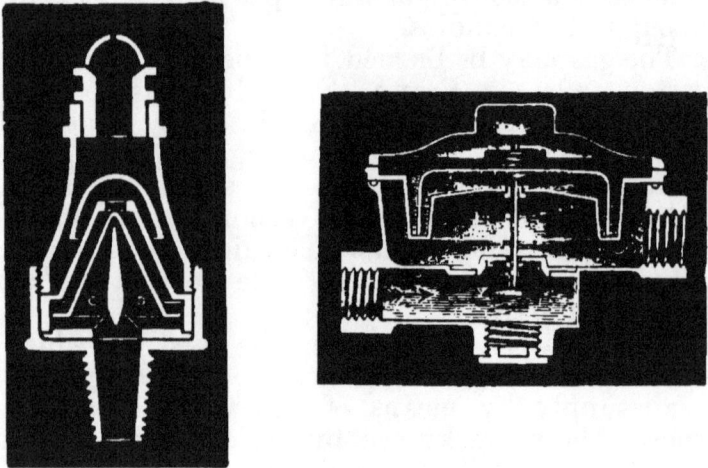

Fig. 7.—Automatic governor Gas-burner, and mercurial governor for main gas-pipe.

were consumed, which on the same basis would bring the required amount of air to 1,700,000 feet. Unfortunately for economy in gas lighting, many of the more economical burners in general use are by no means ornamental, and this may be one of the reasons why private rooms and churches, in which ornament is greatly considered, are still lighted in an antiquated and wasteful style. There is no reason why the better burners should be ugly, and some recently introduced patterns are well fitted for more extended use.

Concert-rooms and theatres will soon be lighted

without exception by electricity, but on account of the expense involved it will be probably a long time before this method is largely used in churches and chapels, the smaller class of private houses, and schools.

The following suggestions are made with a view to reducing the expenditure in gas, as well as to improving the illumination. In private houses it is better to use lamps rather than gas in the sitting-rooms. It is difficult to keep such rooms sweet when continuously occupied, if gas of the usual quality is burned in the usual manner. For a study, an Argand burner with shade gives a pleasant light upon a writing-table, and it is well to have a few brackets for occasional use. There exists some prejudice against gas in bedrooms, but considering that the light is required for a short time only, there is not much objection, if the fittings be sound, and it is extremely convenient in case of illness. The burners should be automatic regulators of the best construction, and the light carefully kept to the minimum required. A few large burners give a more economical light than a large number of small ones. Further details will be given in the chapter which treats of house ventilation.

The lighting of churches by gas is as a rule neither convenient from the point of view of the congregation, nor economical as regards gas consumption. The burners are frequently placed just below the clerestory windows, where they light the roof, but not the floor. The aisles are in shadow, and the light being a long way from the congregation, an unnecessary amount of gas is required. Equally bad is the plan of placing them near the floor on standards : here they dazzle the sight and prevent the preacher seeing his audience, though from the point of view of economy it is preferable to the method first described.

(1) The gas-lights should be kept at such a height as to be well above the preacher's line of sight, but no higher. (2) There should be no lights visible behind the preacher, but a light conveniently placed to illumin-

ate his face and desk. (3) There should be no visible gas-lights at the east end. The large sanctuary standards so common are very trying to the eyes, and do not properly light up the decoration of the altar and reredos. On the other hand an arrangement of candles on the altar is very effective, as they give a soft and pleasant light, very different to that derived from gas-burners. The chancel and east end should be lighted by powerful lights, shaded from the eyes of the congregation after the manner of the foot-lights of a theatre. This may be effected by *placing them against the west wall of the chancel*, or in embrasures on the sides, or behind ornamental screens.

Wherever practicable, groups of *large burners* should be used, furnished with ventilating pipes, which should run into the central shaft by which the church is ventilated. This may not be possible in the nave, as long tubes of the kind are unsightly, but it can generally be carried out with those in the aisles, and with the concealed powerful lights by which the chancel is lighted. "High-power burners," *i. e.* clusters giving a light of 100 to 500 candles, are now easily procurable, being made by Bray, Sugg, and others; they are by no means unsightly, and very economical of gas. In the case of mission-rooms and schools, where appearance is less studied than in a church, these should invariably be used in preference to the star arrangement, or iron hoop studded with small burners, generally adopted by architects, and wherever practicable a ventilating tube should be carried from them into the central exhaust.

In the case of small churches, a single short bracket, carrying a 40-candle burner, placed on the eastern aspect of each pillar, is found to light the building well, without any glare in the eyes of the congregation.

In every case where gas is used, a governor should be placed on the main gas-pipe, and carefully adjusted till there is just sufficient pressure to keep the lights steady. The mercurial governors are probably the

best, and are made by a large number of manufac-
turers, Stott, Shaw, and others.

In very extensive and lofty buildings, a governor
should be placed on each floor. One word on the use
of gas-governors. The light given by a burner of
the ordinary type is most economically obtained at a
definite pressure, generally a low one, say $\frac{6}{10}$ of an
inch (water-gauge). At pressures above this, the light
does slightly increase, but the consumption of gas is
great and wasteful. Gas is supplied to houses at
pressures varying according to the height above the
gas-works and the distance from the supply, but
always at a higher pressure than is required for the
gas-burners. This high pressure is useful for heating
burners, cooking apparatus, and the like, but is wasteful for lighting purposes. Where gas is used for
cooking and heating, a special pipe should be carried
from the main supply close to the meter, and a regulator placed on the pipe which supplies the lighting
burners. A simpler and cheaper method for small
establishments is to light only by the self-regulating
burners referred to above, and a regulator is not then
required on the main pipe.

In estimating the amount of light necessary for a
room, if the walls are not of a very dark colour an
allowance of two candles per square yard will be
ample. If the walls are white, half this will be
sufficient.

"Sun-burners" of the ordinary type (*i. e.* clusters of
small burners placed horizontally) are very wasteful
of gas. Wherever it is possible to place a ventilating
burner, either one of the high-power burners referred
to above, or a cluster of lights with reflector, such as
the "Stott-Thorp" light, or a Regenerative burner
should be used. The Stott-Thorp is a greatly improved form of "sun-burner," furnished with an external ventilating tube, as in the old pattern. It is
much more economical of gas, and when arranged in
the form of a sun-light, by no means unsightly.

Speaking generally, all burners fixed horizontally are wasteful, as they require an increased pressure of gas to make them burn steadily.

No mention has been made of the large amount of heat given off from gas-burners. This is one of the reasons why gas-light is so objectionable in small rooms. When, however, ventilating lights are used, the heat becomes a most valuable agent in assisting the ventilation. Still, even then it is not an unmixed blessing, as when the ventilation of a public building is conducted solely by means of the ceiling lights, the exhaust only takes place when the room is used at night. The ventilation arrangements of a public room should work equally well by day or in the evening. The true function of gas is to serve for cooking purposes, for which it is quite unequalled by any other form of fuel.

CHAPTER VI.

THE MOVEMENT OF THE AIR AND LAWS OF AIR CURRENTS.

THE methods of lighting and that of heating, the absolutely indispensable factors in the production of ventilation, having been determined in accordance with the principles laid down in the last chapter, it is time to consider the means by which the air is to be caused to pass, in adequate quantity, through the building.

It must be clearly understood that the air will not pass, unless some force be brought to bear on it. This does not appear to be generally understood, or we should not find so many buildings bristling with ventilators in which ventilation is conspicuous by its absence.

Excluding mechanical power, brought to bear by means of a fan or bellows arrangement, the forces tending to move the air contained in a room are two,

viz. : (1) the wind ; (2) the difference of temperature
between the outside and inside of the building. Given
adequate communication between the inside of a room
and the external air, the movement of air in the
interior will be directly proportional to the difference
of temperature and the force of the wind. The wind
is a disturbing factor in all ventilation schemes, but is
nevertheless of good service when no more certain
force is available. Wind blowing against a building
enters at the side facing the direction of the blast,
while an exhaust is created on the opposite side. Many
hospital wards are still without any other means of
ventilation, and in moderately warm weather, when a
breeze is stirring, fairly good results are obtained.
When, however, the weather is hot and the air stagnant
this method fails, as it does also in cold weather, when
to open windows or ventilators causes bitter and un-
bearable draughts.

Cold air is heavier than hot. Hot air only rises
because cold takes its place, and no form of inlet has
yet been invented which will reverse the laws of nature,
and cause cold air as it enters to go in any direction
but downwards. Nor are the "exhaust" ventilators
placed on the roof any exception to this law. In a
room furnished only with inlet and outlet valves, without arrangements for warming the air, directly the
weather becomes cold the inlet valves cause unpleasant
draughts and are soon closed. Then, inasmuch as cold
air is heavier than hot, a cold draught descends from
the ventilator above, and this is in turn closed.

The only method which can absolutely be depended
on is propulsion or exhaust by means of a fan. Speaking generally, where there are only a few large rooms
to be kept at the same temperature, the best method is
to propel the air through a proper heating chamber
into the building, but when there are many rooms,
some of which require varying amounts of heat, it is
more convenient to exhaust, admitting the air into
each room from the outside over heating coils.

Fairly good results, however, can be obtained by a chimney or exhaust shaft heated by a fire, hot-water, steam pipes, or gas, or with the boiler chimney passing through it in an iron pipe. This method is easily applicable to houses, where the kitchen chimney can be utilized, and very efficient arrangements of the kind have been made in new houses by building a central exhaust shaft sufficiently near the kitchen chimney to acquire heat from it. But in the case of churches, where a tower or spire is often a principal feature, the formation of an exhaust shaft in the tower is easy, and the most obvious way of securing the removal of the foul air. A shaft should be made in the tower by means of wood, or lath and plaster, and heat applied to the base by means of gas. If there be no tower a turret must be constructed for the purpose, which shall contain the shaft, capped by a cowl. Here, however, we have very little aspirating force except that which is caused by the wind, a quite different arrangement from that existing in a long shaft, which is a powerful motive agency. Still it is often the best which can be done under the circumstances. With regard to the so-called "exhaust ventilators," which are extensively advertised for the removal of foul air from the top of a building, we may say that official experiments have proved that there is very little difference in the exhaust power of the most elaborate contrivance of this kind and a simple tube, capped so that wind cannot blow down it. Wind is so uncertain an agent, that it is not advisable to make any ventilation scheme depend on it. A perfectly still day is the time when greatest change of air is required, and the time when all wind-actuated schemes fail. Still, the plain outlet cowl is occasionally of use when no better method is available, and will be referred to later.

CHAPTER VII.

THE MOVEMENT OF THE AIR AND LAWS OF AIR CURRENTS (*cont.*).

REFERRING to the method of mechanical propulsion of the air by means of a fan, referred to above, we may say that this is the only thoroughly trustworthy method of ensuring a change of air. Mines are ventilated sometimes by means of a heated shaft and sometimes by a fan, but the former method is rapidly giving way to the latter. Fans are of two types : (1) the air propeller, of which the "Blackman fan" (Fig. 8) is perhaps

Fig. 8.—The Blackman Air Propeller.

the best known, and (2) the Pressure fan, such as the "Guibal." The first is an admirable machine for moving large volumes of air at low pressure, *i. e.* where a few large rooms have to be ventilated, and where the air passages are short, wide, and free from bends. Under such description would come concert-rooms, churches, and most theatres. The second type of fan is useful where the rooms to be ventilated are scattered and numerous, and the air channels are long and tortuous, such as large collegiate buildings. For this purpose the "air propeller" is of very little use, and the failure of many costly ventilation schemes may be traced to this mistake having been made. A good fan is comparatively noiseless, and absorbs very little power. It may be worked if necessary by a small-water motor from the town supply, if the pressure is sufficient, and if it be only occasionally used, as in the case of a church, the expense would probably be but small. The best way, however, of actuating a small fan, which it may not be convenient to work by a steam or gas-engine, is by means of an electric motor, connected with the general electric supply, which will soon be available in all large towns. The motor is small, not costly, requires very little attention ; it can be placed in the roof or elsewhere, and can be started in an instant by turning a switch placed in any convenient situation. Under these circumstances it is by far the most convenient mode of ensuring ventilation. Unfortunately, there is at present no good motor which will work with an alternating current.

Let us now consider in a little more detail the means to be taken in connection with a new building, on which a sum of money is to be spent sufficient to ensure adequate sanitary arrangements.

The architect should be advised by a sanitary engineer, skilled and experienced in this kind of work, and a carefully-drawn contract should be entered into with a view to securing the following points.

1. The provision of fresh air in certain amount, dependent on circumstances as stated above, to move at a velocity not exceeding five feet per second.

2. The heating of the building and of the incoming air by such methods as will allow of a uniform temperature being maintained under all circumstances. The question of screening, purifying, and cooling the air

should be also considered, if the funds admit. Screening
should be done in all large towns.

When the work is completed, the due carrying out
of the contract should be ascertained by chemical
analyses of the air, and anemometer measurements of
the air-supply.

The above is not difficult to carry out in the case of
a new building, but where one built in defiance of all
sanitary principles has to be dealt with, something less
must be attempted. Still, even in the latter case, a
good deal can be done if the problem is attacked in a
scientific and business-like manner.

The effect of the wind, and of differences of temperature in affecting changes in the air of buildings, has been
already referred to. It is necessary to say a few words
more on this point, as on a misconception of the laws
governing movements of air depend most of the
numerous ventilation failures which we see around us.
Ignorance of these laws, and a blind trust in some
patent system, have induced a feeling in the mind of
the public that ventilation is impossible, and that the
laws respecting the subject are unknown. Until the
public are more disposed to employ mechanical power,
a large number of buildings must depend on what is
called "natural ventilation" for their sanitary condition.

It cannot be too clearly understood that to arrange
for an air-supply to a building is an engineering
problem of the same nature as the supply to a house
or a town of water, gas, electricity, and the like, only a
rather more difficult problem, since all methods of
effecting it are subject to the disturbing influence of
the wind, and the varying temperatures of the seasons.
The engineer who deals with supplies of water or gas
can depend on fairly constant conditions, but in the
case of the supply of air a much greater elasticity is
required in the methods adopted.

Since the principal agent in natural ventilation is
difference of temperature, we may premise that cold air
is heavier than hot, inasmuch as air expands $\frac{1}{461}$ of its

bulk at 32° Fahr. for every degree Fahrenheit it is heated, so that if the air in a chimney be heated 50° above the external air, it will be increased in bulk nearly $\frac{1}{10}$, and therefore become in proportion lighter than it was before. The colder air outside accordingly, being heavier, presses it from below and causes it to rise, just as to pour water into a glass containing oil causes the latter to rise and float on the top of the water. That is to say, it is solely by the force of gravity that hot air rises. It may be considered that it is quite unnecessary to state this simple fact, but it is not at all uncommon to see gas-stoves fixed with descending chimneys, and it is a popular belief that there is some peculiar property inherent in chimneys by which they are bound to "draw," and that if a hole be made at the top of a building the warm air is bound to find exit.

The velocity with which air rises when heated depends on the difference in temperature between the heated and the colder air, and increases directly as the height of the column. Fig. 9 shows the direction assumed by air admitted through an upright "Tobin" tube, as the result of an actual experiment. It will be seen that cold air projected into a warm room rapidly falls, and that warm air projected into a cold room rises. It is obvious that under usual conditions the "Tobin" tube or vertical inlet ventilator does not perform that which its supporters promise.

It is important to be able to estimate as exactly as possible the amount of movement induced in a given shaft, in order that the required change of air may take place. This involves a certain amount of calculation, and rules for ascertaining the size of shafts appropriate to varying circumstances will be found in a subsequent chapter.

Every endeavour should be made to secure for an outlet something of the nature of a chimney or heated shaft. It is not uncommon to see exit tubes from gas-stoves, or tubes intended to act as outlet ventilators, protruding horizontally from a building. Only

when the wind blows strongly on the opposite side of
the building can these ever act in the way they are
intended.   As a rule, they are either inert, or act as
inlets, admitting cold air, to the discomfort of those
inside the building.

Under the influence of the wind, a side-inlet valve
may, and frequently does, act as an outlet; but a

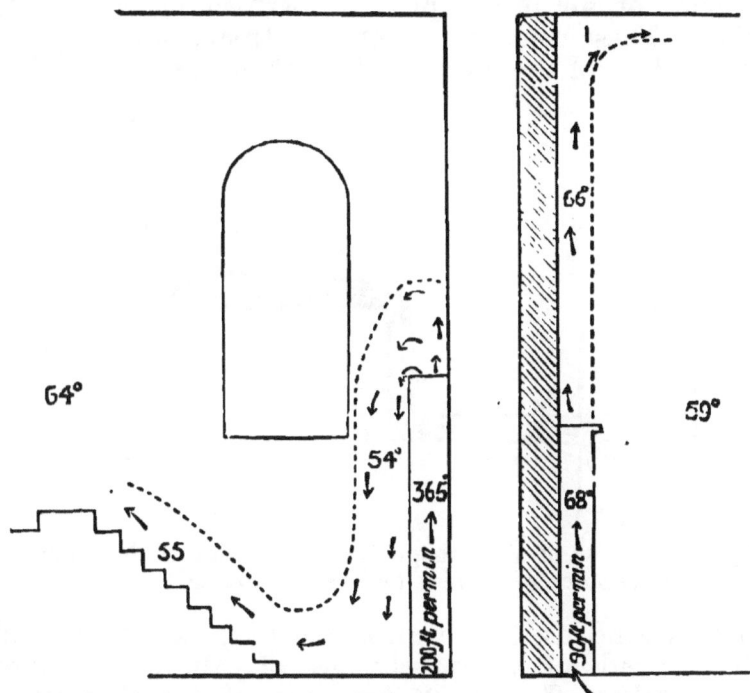

Fig. 9.—Direction taken by current of air admitted through
upright tube at different temperatures.—*Shaw.*

chimney will act not only by the aspirating effect of
the wind which blows across the top, but also by the
difference of temperature, the effect of which may be
approximately calculated by the rules given below.
The passage of air through tubes is governed by laws
which are often imperfectly appreciated.   The dia-

gram (Fig. 10) represents a stream of air entering a
narrow tube, turning a right angle and a rounded
bend, entering a wider passage, reduced again to its
original dimensions, and again emerging into a wide
space. It will be seen from the picture that in every
change of form of the tube, except in the case of the
rounded bend, the stream of air undergoes contraction
—the so-called "*vena contracta*"—and that a certain
amount of air is occupied in performing a kind of
eddy. These contractions of the stream cause a great
retardation of the flow, and it is obvious that all tubes

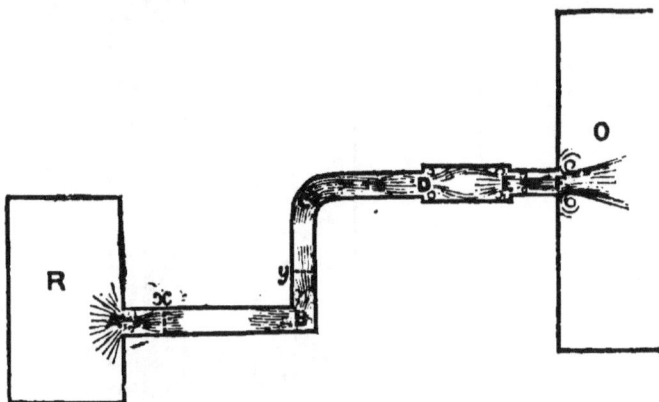

Fig. 10.—Course of air through tube of varying calibre, showing
retardation of current by "*vena contracta.*"—*Shaw.*

of this kind used for ventilation purposes must avoid
sharp bends or abrupt changes of calibre. A right
angle takes off some 25 per cent. of the theoretical
carrying power of the tube. In addition to this, we
have to make a certain allowance for friction, especially if the tube is horizontal. Tubes for ventilation
purposes should be (1) large, (2) smooth inside, (3) free
from sharp bends, (4) as far as possible vertical.

It is no uncommon thing to find air-shafts erected
whose carrying power, owing to neglect of these rules,
is completely nil. Tables will be given in the Appen-

dix by which this friction may be approximately
calculated.

It is bad practice to multiply outlets.   One is better
than many, though it is often convenient to collect the
air from different parts of the room, and run the collect-
ing tubes, which should be as little horizontal as
possible, into a central shaft.   Where several exits are
made to one room it is common to find that, by the un-
equal action of the wind, some of them act as inlets,
causing an objectionable down draught.   No form of
cap or series of deflecting surfaces will prevent a back
current if it be caused by suction, though an attempt
has been made to prevent this by means of valves.
Movable valves are occasionally unavoidable, as in a
chimney breast ventilator when the chimney smokes,
but they should as a rule be eschewed.   They are generally noisy, and frequently stick and become useless.
If the exit has been properly chosen and an adequate
amount of heat given to the shaft, and the supply of
air be sufficient, there will be no down draught.

The following usual modes of arranging outlets are
utterly useless.

1. The exit pipes are led into a system of horizontal
tubes open on each side of the building.   This has either
no effect at all, or more generally causes an objectionable down draught.

2. The exit tubes are led into a false roof, in which
are openings to the air.   The false roof serves as a
great reservoir of cold air, which descends.   The same
may be said of the plan, frequently adopted in
churches, of leaving open spaces under the eaves.   The
only result in cold weather is an objectionable down
draught.

The question of the best position of an outlet from a
room is one frequently discussed, *i. e.* whether it should
be placed at the top of the room, or near the floor, or
at some other point.   Under many circumstances there
will be no hesitation in placing it at the top, especially
if there be much gas burnt without special exit-pipes,

for in the case of a sitting-room, or a church, especially if with galleries, or a theatre, the room gets hotter at the upper part, and this assists the outflow of air. But when there is a well-managed steady flow of air through the room by mechanical agency, it is generally better to remove it at or near the floor level during winter-time. The outlet may be either opposite to the inlet or on the same side. If the air is admitted at a high temperature diffusion takes place better by the latter method. The air being admitted warm rises, and being drawn off below pervades the room more uniformly. In summer, on the other hand, when the air is admitted cool, this is a bad plan, as the cool air will fall directly towards the outlet without being distributed over the upper part of the room.

There is another question which often arises. At what level are the exit-flues from the various storeys to enter the main air-shaft ? There are four ways of effecting this.

1. All flues may enter a shaft in the top storey.
2. All flues may run downwards, and enter the air-chimney at its base.
3. Each flue may enter at its own level.
4. The lower floor flues may be taken downwards and the upper ones at a higher level.

As a matter of construction, it is often inconvenient to run the flues of the upper rooms downwards. On the other hand, there are great conveniences in one central chimney, in spite of the space it occupies. The arrangements for the control of the velocity are easily made, and there is less friction. When all flues enter at the base of the shaft, the risk of one shaft pulling against another is avoided. It is easier to apply heat at the base of the shaft by means of the boiler chimney, or a coil of pipe lining the chimney and heated by steam.

The respective cost of the various systems has been calculated by Planat for a building four storeys high, having a ventilation of 39 cubic feet per second. Planat estimates the cost of coal for heating the shaft

E

by Method 1 at 7 lbs. of coal per hour; Method 2, 4·1 lbs.; Method 3, 5½ lbs. The plan, accordingly, of running all flues into the base of the air-shaft, is not only more convenient but less costly in working, and is the plan generally adopted also when mechanical power is used.

It is hardly necessary to mention the curious notion entertained by some, that because carbonic acid is heavier than air, therefore the foul air tends to fall to the lowest part of a building. On the contrary, the hotter portions of the air are generally the foulest, and they will be found, by reason of their lightness, at the top of the room.

The exit having been determined as regards size and position, the inlets will have to be considered. If cold air be admitted at the ground level, it will give rise to disagreeable draughts about the feet. If it be admitted near the ceiling, it speedily falls, on account of its superior weight, and strikes another vulnerable part of the human frame, the head. Cool air should be admitted at such a height and in such a direction, as to diffuse and mix largely with the contents of the room before it can come in contact with any of the occupants. This is best effected by making the opening about six feet from the floor, and arranging that the discharge shall flow in a direction either vertically upwards, or only slightly inclined from the vertical. This is the only advantage of the "Tobin tube," which, through the energy of its so-called inventor, has largely come into use. So far as it has popularized this principle, the "Tobin tube" has been of service; practically, however, it should never be used, except when it is necessary to take air from the ground level. When an opening is to be made in an outside wall for ventilating purposes, the shorter the path of the air the more effective will the entering be. The aperture into the room should be guarded with a lip, in order to give the air an upward tendency, as in the well-known Sherringham valve. The mouth also is frequently made trumpet-shaped, or containing a series of tubes pointing different ways, as in the "Harding diffuser"; the object of all these devices being to spread the stream of air as it enters, so that it is not felt as a cold draught. At a moderate temperature, *i. e.* not below 45°, a good deal of air can be introduced in this way ; but when the outside air is really cold, a draught will be felt by any one near the inlet. An ingenious addition to the Sherringham valve has been devised by Mr. Bray of Leeds, whereby a series of vanes deflect the air-stream to one side or another, as may be desired. It is important that these inlet contrivances, which we see constantly puffed as panaceas for every kind of venti-lation difficulty, should be estimated at their proper value. They are useful, when large enough, for a certain limited range of temperature, and for certain places, and they have the advantage over window ventilation in that they are not obstructed at night by blinds and curtains. They are often very unsightly, but in a house they are easily concealed behind a picture or mirror, hung so as to lean forwards, and in churches and places of architectural pretensions they may form part of an appropriate bracket. If the air is screened, and this is advisable in private houses, there should be a chamber for the filter, presenting a surface of seven or eight times the area of the inlet, and for this purpose a short upright tube is of service, as a screen can be placed diagonally in it. This, however, will materially check the current, and it must be remembered that the form taken by the stream of air as it enters depends partly on the velocity with which it enters, partly on the form of the inlet, but prin-cipally on the relative temperature, as shown in Fig. 9. A rapid stream of comparatively warm air will shoot high into the room and diffuse widely, a slow stream at low temperature will fall soon and be felt as a cold draught by any one who is near the inlet. When there are inlet-valves on each side of a room, if the wind be blowing towards one side, the valves on

the other side will act as outlets. It is almost needless
to say that unless there be a proper outlet, such as a
chimney, the most "highly recommended" patent inlet
ventilator is quite useless.

When air is properly warmed it may be admitted at
a lower level than when it is cold. It is not good, as a
rule, to admit it at the floor level, though in some cases
it may enter through risers of seats or steps, since if
the room is crowded the air must be admitted some-
what colder than the temperature of the room ; the
same openings may generally be used for both warmed
and cool air.

In calculating the size of inlet openings, we may
assume that a square inch of unobstructed space will
as a maximum admit 125 feet of air per hour, *i. e.* 4
square inches will allow 1000 feet to pass—or enough
for one person if 1000 feet be given per head—if there
be a proper exhaust. Allowing the minimum of 500
feet per head per hour, a room containing 100 people
will, on this calculation, require an inlet space of 200
square inches, or allowing for the friction and the
obstruction of gratings, about 2 square feet.

Inlet ventilators of the kind just described are only
of use in mild weather. When used in a crowded room,
on the first cold day they are closed at the request of
the persons who sit near them, and they generally
remain so throughout the winter. The same rules,
however, apply to apertures through which warm air
is admitted, and as there is little difficulty in making
these do duty both in cold and warm weather, ordinary
inlet ventilators are seldom required in large buildings.

Care must be taken to periodically clean all forms of
inlet valves. "Tobin tubes," as Sir Douglas Galton
remarks, "form very convenient receptacles for dirt,
insects, cobwebs, and dust." Dust filters must be regularly cleansed, or they become worse than useless.

The usual error committed by architects with regard
to inlet-tubes is to make them too small. In most
cases they are mere toys, and would hardly serve to

ventilate a doll's house. The fact cannot be too
strongly insisted on, that under the circumstances of
the low velocities and small motive power available in
natural ventilation, all air-passages must be of a size
which to inexperienced eyes seems excessive.

Another point to be remembered is, that if there is
only one exit shaft (which on the whole is the best
plan), there should be a number of points of entrance,
so that the incoming air should be equally diffused.
Neglect of this precaution has caused some very curious results, notably in the smoking-room of the Reform
Club in London, where, after a fan was introduced for
special ventilation of the room, it was found that the
atmosphere was worse than before. Examination
showed that the current of air passed straight from the
inlet to the outlet in a single stream, leaving the rest
of the room stagnant. This can be easily remedied by
dividing the entering stream of the air by means of a
simple diffusing apparatus.

The question of diffusion has not been referred to
with reference to ventilation. That a considerable
amount of change of air occurs by simple diffusion
through walls no one can doubt. Indeed those who
live in slightly-built houses in windy situations, find
that they get far more air in this way than is pleasant.
The amount, however, thus received is so dependent on
the nature of the situation of the building and the
force of the wind, that the consideration need not
affect our calculations.

External air grates must be made of such size as
to give an unobstructed area equal to that of the
opening in the wall, *i. e.* 25 to 50 per cent. longer than
the wall aperture.

We shall now give a sketch of the methods of ventilation most applicable to different classes of buildings.

CHAPTER VIII.

THE VENTILATION OF HOUSES.

HAVING laid down the general principles on which
the air-supply to dwellings, &c. should be arranged, it
will be convenient to discuss their application to
specific classes of buildings.

The Englishman is exclusive. He loves not to live
in large hotels like the American, and is only just
beginning to tolerate flats. His house is his castle,
and he keeps it tight and impervious. The plumber is
privileged to lay on sewer-gas from the main, but no
arrangements are made for the provision of fresh air.
One redeeming feature there is. The open fire is a
peculiarly English institution, and an open fire means
a wide open chimney, almost the best extract ventilator
possible.

In consequence, the average English house is much
less stuffy than similar dwellings in other temperate
climates, and if it were not for the large amount of gas
burnt for lighting purposes, very few special ventilating
arrangements would be necessary.

Nevertheless, many English houses are, in certain
respects, much less comfortable than they might be.
In the ordinary town house, with kitchen in the basement, the odour of cooking three times a day pervades
the house, for the largest ventilation outlet in the
kitchen is the staircase which leads to the dining-room.
In cold weather windows and doors are closed, and fires
are kept up in the sitting-rooms. The powerful suction
of many chimneys draws air from every quarter—from
basements, with perhaps decaying refuse, from dusty
coal-cellars, from water-closets, and other undesir-
able places, which are seldom completely isolated.
There is not often a central heating apparatus, and the
passages are very cold. In the sitting-rooms a blazing
fire makes an area of warmth near the fireplace, but it

is impossible to sit near the windows on account of the piercing draughts which penetrate blinds and curtains. One steps from a sitting-room at 62°, through a passage with a temperature of 40°, into a bedroom often at a lower temperature still, or badly warmed by a recently-lighted fire ; pleasant enough, it is true, to look at, but giving a minimum of warmth with a maximum of trouble.

Now the ventilation of an ordinary dwelling-house is an easy and not very costly business. Of exhausts there are plenty, in the shape of chimneys. What is wanted is a supply of pure warmed air.

Every house should have a heating apparatus connected with an air-supply, by which a stream of warmed air can be continually poured into the building. This may be done in many ways. In small houses probably the best method is by a stove in the central hall or entry, furnished with a duct communicating with the outside. In large houses a hot-water coil is more convenient and less unsightly, standing over or in front of a similar air inlet. There should be an inlet into every room above the door, concealed by a projecting lintel, and the warmed air will find its way, drawn by the various chimneys, into every room. This is a very efficient and by no means costly way of effecting both warming and ventilation.

The plan to be adopted with an ordinary-sized sitting-room must depend greatly on the number of people who inhabit it, and the question whether gas is burned. The first thing which requires attention is the fire-grate. This is too often constructed mostly of iron, with chimney at the back, giving, it is true, a bright blazing fire, but drawing cold draughts from all the windows and doors. In vain windows are pasted up and india-rubber put round doors. It is the old story of trying "to expel nature with a fork." If there be an adequate supply of warmed air to the room from some other source, such as a hot-water coil furnished with inlet, there is not much objection to the quick combustion grate, except that it burns an unnecessary amount of coal. Under most circumstances, however, a grate of tolerably slow combustion is preferable, like those of the modified "Rumford" pattern, made by the Teale Fireplace Company, or the "Marlborough," with closed ashpits, or the "Abbotsford," "Waverley," and others, with solid brick base. In the former of these the chimney draught is considerably checked, but there is enough to keep the room sweet if there be but few occupants, and no gas is burned. Speaking generally, however, a grate of the "Rumford" type, with closed ashpit, seems to be the best for an ordinary sitting-room. If the room be habitually rather crowded, it may be better to have the chimney opening made larger than is usually provided.

If there be an air-supply from the hall over the door in the way suggested above, no ventilator or additional inlet will be required, provided the windows can be arranged to open without causing much draught. Sash windows are the most convenient in this respect, as they may be kept slightly open at the top, or better, provided with a board fixed against the sill, so that the lower sash may be raised about four inches without the air being admitted below. The opening between the sashes then admits air in an upward direction as required. It may be occasionally advisable to insert an inlet-valve in some convenient place, *i. e.* where the draught, which it will unavoidably cause, will not traverse the usual sitting-places in the room. A "hot-air grate" renders inlet-valves quite unnecessary, and if the room is very large, it will be found that an ordinary grate is insufficient to warm and ventilate it. Either a hot-air grate, or a hot-water coil in conjunction with an ordinary grate, should be used, the radiator being furnished with an inlet opening. Great trouble is caused in houses by the black marks which appear above pipes used for heating. This is caused by the impinging of the stream of warm air with its dust particles against the wall. Hot-air grates should be made with the air-flues delivering through a perforated or fretwork back, in a panel of an over-mantel of dark wood (*vide* Fig. 3, p. 38). Hot-water coils should have ornamental marble tops, or shelves fitted over them, if they are close to the wall.

The kitchen is generally a difficult place to ventilate. Cooks do not like draughts, and exits into the chimney are apt to spoil the draught of the range. There would be no difficulty in making a cooking-range which should introduce warm air, but there are very few of this kind in the market. Probably a small chimney-breast ventilator and an inlet valve would be the best method. But a great deal can be done by avoiding unusual sources of air fouling. All gas-stoves should be fixed near the range, with a large hood over them ventilated into the chimney. When a cold kettle is placed over a cooking (Bunsen) flame, the latter is cooled so that the gas is improperly burned, and a strong-smelling and deleterious gas called "acetylene" is formed. There are many houses where this odour can be perceived all over the building when the gas cooker is being used. Special gas-stoves are not necessary, unless the existing range is too small. There is no difficulty in adapting gas to an ordinary cooking-range, so that either gas or coal may be used, as is found convenient. Boiling burners may be placed on the hot-plate, and a special gas-burner placed to heat the ovens, at a very small cost. The fumes of the gas, of course, escape by the chimney.

Bedrooms will not require much special treatment if there is a central warming apparatus. In case of illness, however, a slow combustion grate, which will keep alight without attention all night, is a great convenience. Of late, gas fires have become very popular. They are certainly very convenient, giving mostly radiant heat, and that in a very short time after they are lighted. There are certain objections to them. First, they are sometimes fitted in chimneys in which there is a down draught. If the chimney smokes with coal, it is likely to smoke also with a gas fire, and the fumes from the latter are nearly as unpleasant and unwholesome as from a coal fire. Secondly, in fixing a gas fire, the plumber invariably partially stops the chimney opening with a piece of sheet-iron. This renders the chimney almost useless as a ventilating exhaust, though it greatly economizes the heat from the fire. Probably the introduction of a chimney-breast ventilator near the ceiling would get over the difficulty.

The billiard-room will require special attention, as there is need of a strong light and evenly-distributed heat, and smoking generally takes place in it. It is best heated by a fire and hot-water pipes. Inlets may be made behind the hot-water radiators, and the fire should have a rather quick combustion, while a chimney-breast ventilator may be fitted near the ceiling. It is well, however, to have a special outlet shaft, which may run alongside of the chimney, for use in summer, and ventilating-pipes from the gas-burners should be carried into this.

The table should be lighted by Regenerative or incandescent gas-lamps, furnished with exit-tubes, the ordinary arrangement of three small burners projecting from the interior of a ring is unsteady, and very wasteful of gas.

Gas as an ordinary illuminant in private houses is an objectionable method of lighting. The fittings as a rule are unsightly, and the fumes are unpleasant to the senses, and corrosive of leather and metal-work. When a room is lighted by gas the currents of air arrange themselves in a rather curious fashion (*vide* diagram, Fig. 11), there being three strata of air of different temperatures. The gas-burners keep up a small local circulation of heated air at the upper part of the room, as shown in the diagram. This does not, as a rule, extend much below the level of the burners, which are generally placed above the level of the head. Next there is a temperate zone in the centre, and a cold zone for about a foot above the floor. At the same time the upper and middle zones are always mixing by a process of diffusion. This explains how intolerable the atmosphere is in low rooms lighted by gas ; the gas-burners being necessarily placed lower than usual, the heads of the occupants come within the hot zone near the ceiling.

It is convenient to have one unobtrusive gas-bracket in every room, for the convenience of the servant who cleans the room. This will avoid the necessity of

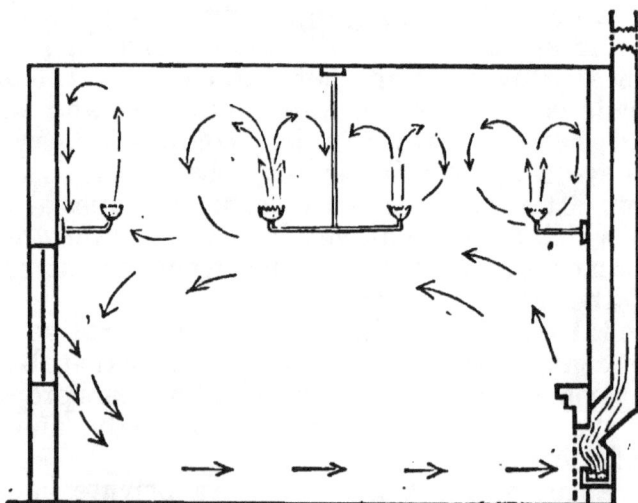

Fig. 11.—Circulation of air in room lighted by gas.—*Shaw.*

carrying about lamps and candles, with the dangers attending their use. If a ventilating-tube can be run from the ceiling into the chimney, a Regenerative light is a convenient method of lighting a library or dining-room, and a chimney-breast ventilator should always be used, whether gas is burned or not. If the use of gas is unavoidable in the sitting-rooms, the burners should be of the self-regulating type, with globes widely open at the base, and the light carefully apportioned to the size of the room according to the rules given above, brackets from the walls being generally more convenient than a centre pendant.

Perhaps the most severe test which can be made of the ventilation arrangements of a house is on the occasion of an evening party. On a festival of this kind it is an exception to find an atmosphere other than very foul. The rooms are crowded with a number of guests, and extra light is required, which means increased heat and impurity of atmosphere. The following hints may be useful.

1. Lights used exclusively for decorative purposes should be small. It is common to find a powerful lamp of, say, 40-candle power covered with a thick shade, reducing the light to about 5 candles. It would be better to have a 10-candle lamp with a less opaque shade.

2. Let the rooms be thoroughly warmed before the guests arrive, and then reduce the fires to the smallest dimensions.

3. It is generally possible to keep a number of windows slightly open, if they are shielded by curtains, and it is not suspected that they are open. A screen of muslin placed in an open window may be used to check the draught.

4. Any room used for dancing should have the windows widely opened between the dances, while the sitting-rooms may be aired in a similar way when the guests are absent at supper, &c.

It sometimes happens that a house not specially built for the purpose has some parts of it used for purposes of a school, either regularly or occasionally. It is common in vicarage houses to have a special room, "The Parish Room," fitted up for confirmation classes, Sunday-school, and the like. These rooms seldom have any ventilation arrangements, and are generally very offensive when used. It is best here to warm by a hot-air grate, and light (if by gas) by a central single burner with reflector and ventilating-tube, which passes horizontally from the ceiling into the chimney. If there be no gas a chimney-breast ventilator will be the simplest arrangement. Sash windows should be utilized according to the method described above, and other windows should be fitted with hoppers 'so as to admit the stream of air vertically. The same treatment may be adopted in smoking-rooms if they are of small size.

One room in the house must not be forgotten, viz. the water-closet. This is frequently a source of great danger to the household, in spite of the fact that usually at the present day the soil-pipe is carried up to the roof, and an efficient trap provided. The ideal arrangement is that invariably, when possible, carried out in hospitals, where sanitary annexes, divided from the main wards by ventilated lobbies, contain all that is likely to become offensive. The water-closet should have two openings, which cannot be closed. A single air-brick or single small window is of very little use. Probably the best arrangement is an inlet ventilator, arranged so as not to cause a draught near the occupant, and a shaft carried from the ceiling up to the roof, capped by a cowl.

In making ventilation schemes for houses it is very necessary to regard people's natural prejudices. We all know the amateur ventilator, whose house is a perfect whirlwind of cross draughts, which he delights to show to visitors. We can, on the other hand, be too scientific, and there are cases on record where the most perfect and carefully-devised schemes have come to nought because they involved the absence of an open fire, and perpetually-closed windows.

The Ventilation of Cottages.

Model cottages are generally furnished with an ample supply of air-bricks and ventilators of various kinds. The first thing the occupant does is to carefully stop all these apertures with paper, old rags, or anything that comes handy.   As a rule, he is economical of fuel, except in the coal-producing districts, and though he may work most of the time in the open air he dislikes draughts.

The open range which is generally used for cooking in cottages will prevent the living-room from becoming very offensive.   The only measures for ventilation which are likely to be tolerated by the ordinary occupant of cottage property are—(1) the insertion of a simple chimney-breast ventilator into the chimney of the cooking-range, and (2) openings may be made over the doors of the bedrooms.   The latter will admit air, more or less warmed by the fire below, and this will escape by the bedroom fireplace.

It would be easy to make the ordinary open range into a hot-air grate, and this would be by far the most economical plan as regards heat and ventilation. Messrs. Shorland of Manchester make a chamber which can be fitted at the back of any grate, by which air may be admitted from the outside, and delivered into the room warmed.

CHAPTER IX.

THE VENTILATION OF CHURCHES.

CHURCHES being occupied for comparatively short periods, a smaller allowance of air is permissible than in houses, schools, hospitals, &c., but at least 500 feet per hour should be supplied for every person and every foot of gas which is burned.   As persons attending churches are dressed for walking, a lower temperature is permissible than in concert-rooms and theatres, in which a number of people are in evening dress.

The air movement may be either by a shaft in the tower, or by a fan.   So many recently-built large churches have their organs blown by a gas-engine that it would make very little difference in their arrangements were the engine-room built sufficiently large to accommodate an engine and fan for ventilation purposes. The air should be propelled into the building through a chamber filled with hot-water pipes, arranged in sections, so that the heat may be entirely under control. It should be delivered at a number of different points, and apertures made near the roof by which it may escape.

A certain amount of hot piping may be distributed in the church, especially near the doors, and these pipes should be kept hot the whole week. They will serve to prevent the building from cooling, and to make it comfortable for small congregations at the week-day services. The precise amount of heating surface necessary to maintain the temperature may be ascertained from the rules given in Chap. XVI. This is the best method of ventilation for a church, but if the scheme includes a lofty tower, part of this may be enclosed to form an exhaust shaft, and the shaft itself heated by gas or by a coke furnace.

The admission of the warmed air will then have to be carefully considered. The coils may be placed either enclosed in cases against the walls of the church, with apertures behind, or in culverts below the floor, communicating with the external air by openings of proper size, and with the interior by tubes built into the walls, terminating in trumpet-shaped openings about seven feet from the floor. These openings should be made sightly by appropriate architectural treatment. The openings should be arranged in well-distributed positions, mainly in the part of the church most remote from the exhaust shaft, so that the air may pass uniformly through the building. Fig. 12 gives a general arrangement, suitable for a small church. In a large building the warm air should be admitted at a greater number of points.

For summer ventilation the shaft will require a

larger amount of heat, and air may be admitted also through windows at a tolerably low level.

It is advisable, in order to prevent the cold down draught which is so commonly felt in churches during the cold weather, that the clerestory windows should be of double glass, and the roof made as impervious to heat as possible by means of felting and boarding.

In churches with roofs not entirely open, it is convenient to run a horizontal shaft along the roof, com-

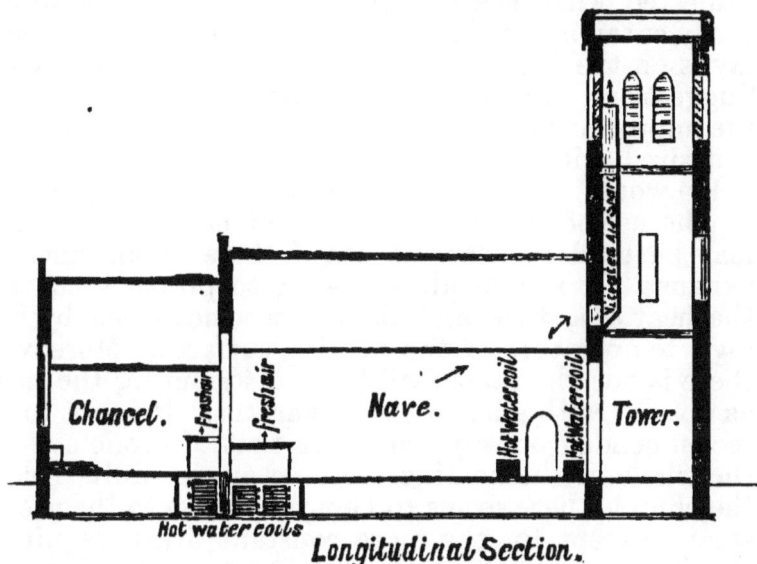

Fig. 12.—General arrangement of heating apparatus and exit-flue for small church.

municating with the church by a number of apertures. This should be led into the main exhaust shaft, and it will be convenient, if ventilating gas-burners are used, that the exit-pipes from the latter be led into the horizontal shaft.

The consumption of gas should be carefully checked by the rules laid down in the chapter on gas-lighting. It is not generally known how much an organ is affected by difference of temperature. Ten degrees of tempera-

ture above that at which the organ is tuned will serve
to introduce the most horrible discord in a perfectly-
tuned instrument. This may be often noticed at an
evening service. In a properly-ventilated church the
temperature should not rise.

Revolutionary as these proposals may seem to one
accustomed to the conventional (and insanitary)
methods of church construction, they involve very little
alteration in the usual structural arrangements, and if
projected when the church is planned, do not necessarily entail a very large expense. The principal outlay is on the increased amount of heating apparatus.
Suggestions are given in a subsequent chapter as to
the improvement of existing churches where ventilation
is desired, with regard to other methods of heating.

We would refer to the remarks made in Chapter IV.
on the use of "hot-air" methods of heating. A stove
may be used to heat the air, if there be an efficient
exhaust. It is difficult, however, to properly control
the heat of a stove, and the air becomes dried by the
high temperature of the heating surfaces. Moreover,
there is not the same facility in distributing the heat
as obtains with a hot-water apparatus. It is not to be
recommended on any score but that of economy. If
the whole of the heating surfaces are in vaults below
the church, there seems to be no objection to the use of
steam, except for the more constant firing required.
But if the radiators are in the church, it is difficult to
make the system at all times silent.

CHAPTER X.

THE VENTILATION OF SCHOOLS.

THE difficulty of ventilating schools increases directly
with their size and complexity. A large collegiate building, such as one of the University Colleges which have

F

arisen in our largest towns, consists of a number of rooms
of various sizes, some being sitting- or common-rooms,
others class-rooms, generally crowded, while there are
besides chemical laboratories, in which special arrangements are required. School buildings of the more
complex type present the most difficult problems to
the ventilating engineer, and it is impossible in a work
like the present to do more than point out some of the
more important points which have to be considered.

The principal failures in the ventilation of large
buildings of this type have been through attempting
to ventilate too large an area by means of one fan or
other exhaust. This results in the rooms near the fan
becoming efficiently ventilated, but the far ones receive but little benefit. It will generally be found
convenient to have a special fan or chimney exhaust
for the chemical laboratories.

It is no less important that the air-passages should
receive careful attention, not only as to their design,
but in the course of their construction. The allowance
for friction, for length and angles, is generally under-
estimated.

In Mr. Robins' valuable work on technical school and
college building, a careful survey is given of the ventilation arrangements in a number of recently-erected
colleges and technical schools of large size.

There are three methods by which air may be
supplied, viz.—

1. Supply from one large heating chamber.
2. „   „   separate heating chambers.
3. „   „   cased coils in the rooms.

For colleges, and similar buildings containing many
rooms, in which various temperatures are required, the
first method is inadmissible. A room with a southern
aspect and many students receives air the same
temperature as a north room with few occupants.
Under these circumstances, the first will be too hot
or the second too cold. Reducing the heat can only be
done by reducing the air-supply. The difficulty can,

however, be got over by having a number of auxiliary heaters attached to each room, which can either be used or not, and the propulsion system is then available. Whatever system is used, it is necessary to have a separate heating surface to each room.

Mr. Robins' review establishes the great advantage of extraction by fan, and though an exhaust chimney, heated by the flue of the boiler passing through it, is sometimes the only practicable arrangement, the amount of heat communicated to the shaft is not as great as calculation would lead one to imagine. In several instances it has proved insufficient, and a mechanical arrangement has had to be adopted for special rooms. In Chapter X. a description is given of a moderate-sized school, heated and ventilated by means of steam pipes.

The question of who shall be responsible for the condition of the various rooms in a large school is one which frequently arises. The plan is adopted in some establishments of giving the masters or lecturers a key with which the heat can be controlled. This acts badly where the same room is used by various lecturers, whose tastes on the subject of temperature may be various; but when one master remains in the same room for several hours, it is open to less objection. On the whole, it is better to have a caretaker, who will adjust the apparatus, if one of sufficient intelligence can be found. He should visit the rooms periodically, and be responsible, (1st) for the rooms being sufficiently heated in the morning before the school opens; (2nd) for starting the ventilation apparatus, and adjusting the mixing valves or other mechanism provided, so that the incoming air should be of the right temperature. What this temperature should be must be determined by experience. In a room scantily filled it may require to be many degrees warmer than the desired temperature of the room; if the room be well filled, considerably lower. Of course thermometers should be placed in every room, and the place for these

should be carefully chosen. They should, if possible,
not be hung against the wall, which will probably be
colder than the air in the room, nor anywhere near the
heating apparatus.

Any lecture-room which is but seldom used should

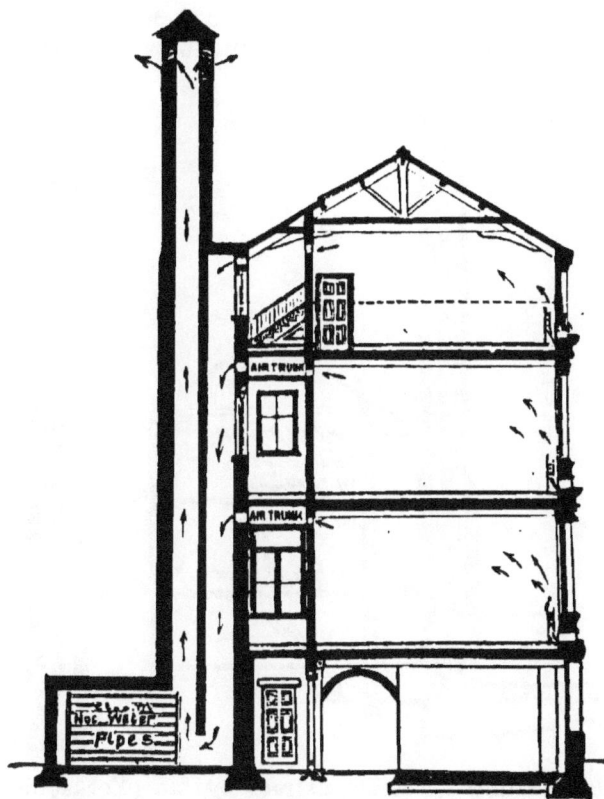

Fig. 13.—Extract and heating by steam (Leeds Boys' Modern School).

have its outlet opening almost closed, except when it
is used. By this means the ventilation arrangements
of the other rooms will be more effective. With regard
to chemical laboratories, an ample supply of draught
chambers, and the enforcement of strict rules as to

their use, will greatly relieve the ventilation requirements. In the Finsbury Technical College, the Bunsen
burners at the working-benches are fixed under an
exhaust hood, and they cannot be used in other
positions.

Fig. 13 shows the method of ventilation and heating
of a block of buildings at the Leeds Boys' Modern
School. The upcast exhaust-shaft is heated by steam

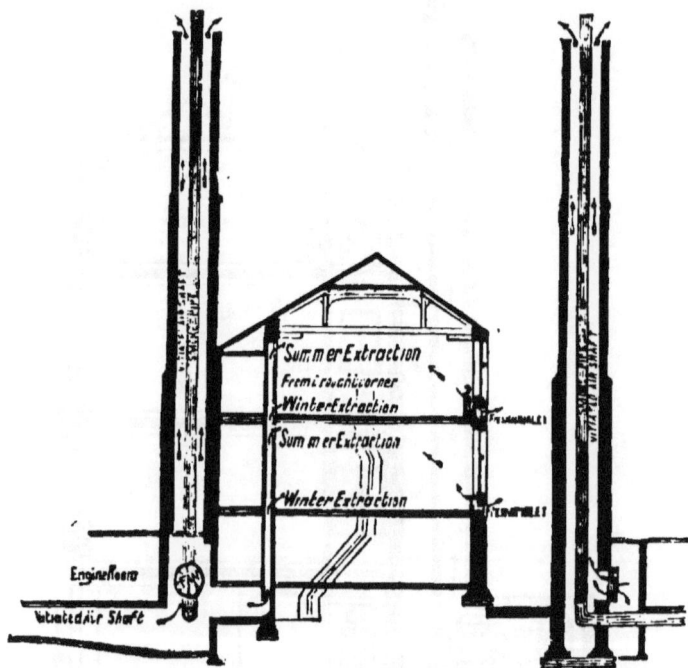

Fig. 14.—Heating by steam, extract by fan (Yorkshire College,
Medical Department).

pipes in a chamber at the base, and the air is drawn
down from the upper parts of the rooms. The air is
admitted over steam coils placed against the walls,
provided with the casing and mixing valve shown at
Fig. 5. By this means the air can be admitted at
any desired temperature. Fig. 14 shows a section of

the newly-erected Medical Department of the Yorkshire College, Leeds.   The air is admitted over steam coils as in the last example, but the exhaust is by a gas-engine, which actuates a "pressure" fan.   The air is removed from the rooms at the lower part in the winter, and at the upper part in summer.   The building was designed by Mr. W. H. Thorp, of Leeds, and Messrs. Ashwell and Nesbit of London undertook the ventilation and heating.

Elementary schools in large towns are generally extensive enough to make it worth while to supply a proper scheme of mechanical ventilation.   Consisting as these buildings generally do of a limited number of large class-rooms, the propulsion system here works well, and is strongly recommended by Dr. Carnelley in his report to the Dundee School Board.   Dr. Carnelley's recommendations at the conclusion of his report are worth quoting in abstract, being as they are the result of a very complete examination of the question by a scientific man well qualified to judge.

Recommendations.

I. For new schools, mechanical ventilation should be employed.

a. Because it is more comfortable, the temperature even, and draughts avoided.

β. It is healthier, and prevents the spread of infectious diseases.

γ. It increases the working and therefore the grant-earning power of the children, and the teaching power of the staff.

The extra annual cost of mechanical ventilation is about £39 a year for a school of 1000 children, or $9\frac{1}{2}d.$ per head.

II. A gas-engine (about 2 h.p. as a rule) should be used to drive the fan.

III. A 48-inch fan should be used to propel and not exhaust the air.

IV. High-pressure hot-water pipes should be used for heating, the pipes being placed in the air-chamber, and not in the flues.

V. Arrangements should be made for mixing cold air with the hot, before the latter enters the room.

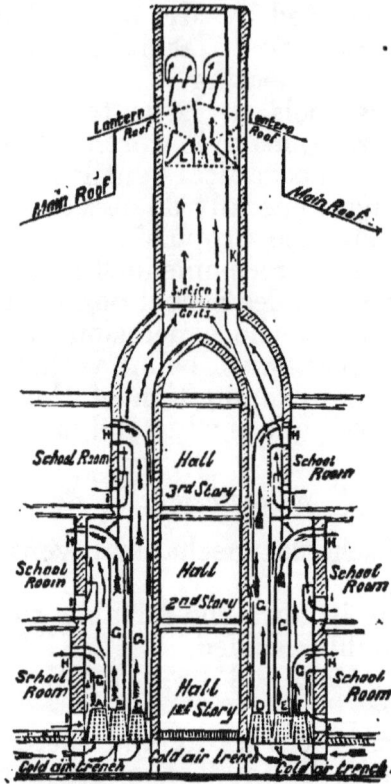

Fig. 15.—Ventilation of school. Steam extract and heating.
(*Billings.*)

(The same end may be obtained by arranging the hot-water pipes in sections.)

VI. The air should be filtered through coarse jute cloth, placed diagonally across the large inlet flue, or across the air-chamber.

A special form of brick furnace is also described, applicable to high-pressure hot water.

For small village schools or class-rooms the simplest method is by one or more hot-air grates, the air-passages being made of ample size, and arranged to deliver about six feet from the floor. An exhaust-shaft should be provided, in the form of a turret on the roof, with an exhaust-cowl, or a shaft running up parallel with the chimney.

Sometimes a village school has a cottage for the mistress or caretaker attached. In this case the shaft may be made to adjoin the kitchen chimney of the dwelling-house, and will be warmed by the constantly burning fire in the latter.

The windows should open with hoppers.

CHAPTER XI.

THE VENTILATION OF HOSPITALS.

THE provision of an equable flow of warmed air through the wards of a hospital is by no means an easy problem. The rooms are occupied continuously day and night, and there are frequently more or less offensive odours to be carried off. It is generally agreed that not less than 3000 feet of air per head per hour should be given. Moreover, the usual shape of a hospital ward, a long oblong, has been fixed on mainly for the convenience of cross ventilation by windows on opposite sides, and does not lend itself readily to systems of mechanical ventilation. There are probably few buildings for which so many special systems of ventilation have been devised, or in which there have been so many failures. Hospital wards in England are generally sweet and free from offensive smells, but this purity of atmosphere is gained at the cost of numerous and unpleasant cold draughts during the colder part of the year, which patients greatly resent. "We have a regular fight with the patients about the windows," said a nurse in one of our large hospitals recently to the writer, and indeed the sufferers can hardly

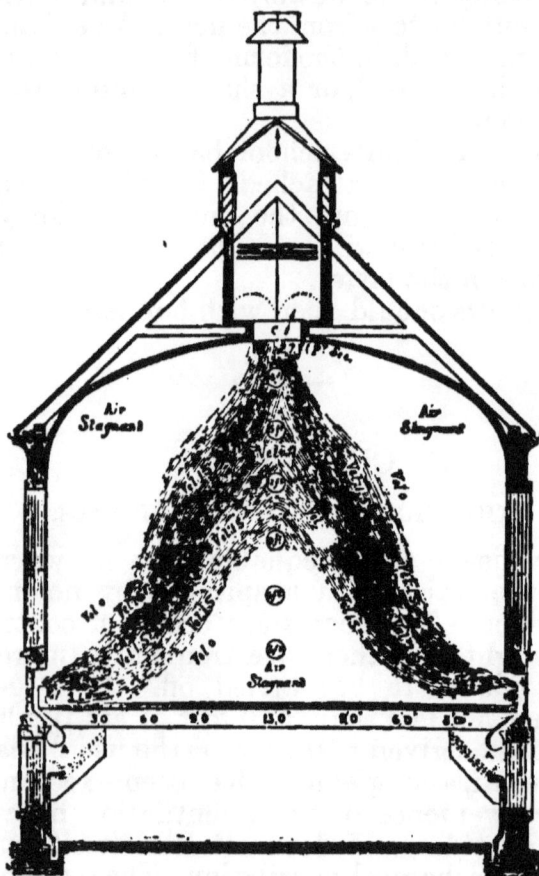

Fig. 16.—Currents of air in hospital ward.—*Galton.*

be blamed. It should be possible to keep every hospital ward of whatever size perfectly sweet without the necessity of opening a single window, but it is very seldom that this can be done. No doubt the problem

is a difficult one with a long ward, though comparatively easy in the case of the circular wards which
seem to be coming into use. The diagram (Fig. 16)
shows that even in a mechanically-ventilated ward
there may be a stagnant area in the centre, though this
could be considerably improved by making the ward
lower, and fitting a more perfect diffusing arrangement to the inlets.

Among the methods in use for hospitals may be
mentioned that adopted at Victoria Park Hospital for
Diseases of the Chest. Here a central exhaust tower,
heated by hot-water pipes, draws air down a shaft some
distance from the building into a double culvert in the
basement, which runs the whole length of the building.
One of these channels contains hot-water pipes, and
the two are made to communicate by means of swing-
doors, so that the hot and cold air may be mixed. The
air rises through vertical pipes into the wards, and is
delivered through gratings about three feet from the
floor, and exhausted from openings just below the
ceiling in a line above the inlet. On the whole the
system works well, except in rather warm and close
weather, when the difference of temperature within
and without is not sufficient to cause an adequate
draught in the exhaust-shaft. The inlets opening
horizontally into the room cause a very perceptible
draught to those near, and patients accordingly are
prone to stop the openings with pillows or anything
that is available. It would be better if the incoming
stream of air were better diffused on entering, and in
a part of the ward more remote from the exhaust.
A very complete and successful system of mechanical
ventilation is in use at the recently-constructed
Victoria Hospital at Glasgow, where the propulsion
system is used. The air is screened through a
specially-constructed filter, designed by the engineers
(Messrs. Key & Tindall) who carried out the work.
The screen is kept constantly wet, and the air washed,
not only the solid particles being removed, but also fog.

The air is driven over heating coils, and admitted into the wards at the ceiling level, being removed at the floor level.

There is some difficulty, however, in forcing air by propulsion to the end of a long ward. The exhaust system again is open to the objection that if a window near the outlet be open, most of the air will be drawn from the window, instead of from the end of the ward. Again, the ventilation being required equally day and night, the fan must be constantly working, involving constant attention to the engine. This can easily be effected in a large establishment, but is difficult in a small one. The alternative is to furnish each ward with a separate exhaust, heated by gas or steam, or to take all the flues to a central chimney. Steam is often used to heat hospital wards, and then the coils should be furnished with the mixing-valve arrangement described in the chapter on warming. The heating of the ward is preferably done by open fireplaces, the steam being used only to warm the incoming air. Wards heated entirely by steam are apt to be oppressively hot and stuffy. The modern practice of building hospitals for infectious diseases in the form of isolated one-storeyed pavilions makes a central system impossible. There should be no difficulty in contriving an exhaust-shaft, heated by the fire of the kitchen range, which always forms part of the fittings of the pavilion.

Small wards should have hot-air grates, with air-passages of ample size, large chimneys, and a chimney-breast ventilator. The windows should open inwards, on the hopper principle, but arrangement should be made to admit sufficient warmed air to render it unnecessary to open the windows.

In connection with hospitals, the waiting-hall for out-patients requires very complete ventilation, and among the examples in a subsequent chapter will be found a method applicable to such cases. But the ventilation may be carried out on the general principles laid down with reference to assembly-rooms.

In the very extensive out-patient department recently erected at the General Infirmary at Leeds, the large waiting-hall is supplied with warm air by apertures beneath open steam coils, the smaller (examination) rooms are furnished with hot-air grates, and the larger (consulting) rooms are heated by open fireplaces, but there is a large inlet aperture, guarded by a steam coil with case and mixing-valve. The large hall has three roof turrets, which can be heated by gas, and the system on the whole is found to work well.

The dispensary at a hospital is often the source of strange odours. The dispenser should be furnished with draught cupboards, like those in a chemical laboratory. The room should have an ample supply of warmed air, and a special exit.

The following is the method of ventilation in use in military hospitals, as arranged by the Barrack Improvement Commissioners, and the same plan is adopted also for the ordinary barrack-rooms. The size of the openings given are those in use in barrack-rooms ; those in the hospitals are nearly double the size.

1. A smooth, straight, wooden outlet-shaft is carried from the ceiling to six feet above the roof. This is not larger than one square foot in area. If a larger is required, two or more are put up. Ten to twelve square inches are allowed for each man, according to the floor on which the room may be situated.

2. Sherringham valves are fixed about nine feet from the floor, an inlet space of rather more than a square inch being allowed for every sixty feet of space in the room. The room is warmed by a "Galton" hot-air grate, which affords both an ingress of warm air and an outlet by the chimney.

This system is said by Dr. Parkes to "work extremely well," but the author's experience is that the Sherringham valves are always closed in cold weather, and the inlet space is then not sufficient,

CHAPTER XII.

THE VENTILATION OF THEATRES AND CONCERT-ROOMS.

THE adequate supply of fresh air to so large- and elaborate a building as a theatre is a matter of such magnitude that an expert is likely to be consulted on the matter. There are few more perfect schemes of ventilation than have been devised and are now satisfactorily working at certain theatres. Notably, the Opera Houses at Vienna and Frankfort are instances of the kind. The Madison Square Theatre in New York is one in which the cooling of the air is very effectively done. There are also several theatres recently erected in London where mechanical ventilation is effectively performed. The Criterion Theatre in London is entirely below the level of the street, and has therefore to be ventilated on the same principles as a mine.

The arrangements for the heating and ventilation of the Vienna Opera House are singularly complete. They were designed by Dr. Böhm, the medical director of the Hospital Rudolfsstiftung. There are two fans, one for propulsion and the other for exhaust. The air is heated by steam coils, and is admitted by the floor and through the risers of the seats. Each gallery and compartment, including the stage, has its own independent supply and means of heating. The velocity of the entrance of the air is between one and two feet per second. The air is admitted to a basement chamber, into which, in summer, sprays of water are introduced ; it is then driven over the steam piping, and on into a mixing-chamber. The whole of the heating and the air-supply can be controlled from one central office, and by means of electric thermometers the officer on duty can ascertain the temperature of every part of the house.

Very similar arrangements are found at the Metropolitan Opera House, New York ; but there is but one

fan, and that is used on the " plenum " or propulsion
principle, which aims at producing a higher pressure
inside the house than outside. This is designed to
avoid draughts from the doors, which are so usual in
theatres ventilated on the exhaust principle. The
heating arrangements are excellently managed. The
temperature of the house is kept at a high point during
the day, with slow movement of the air, in order to
warm the walls ; but when the house is about to be
occupied, the temperature is dropped to that at which
it is desirable the building should be maintained. As
the performance goes on, and the warming action of
the occupants becomes felt, the temperature of the
incoming air is gradually lowered. There is found to
be remarkably little difference between the temperature
in various parts of the house, and there are special
arrangements for regulating the hygrometric condition
of the air.

In former times there was great difficulty experienced in preventing the fumes from the foot-lights
contaminating the air of the house, and few of the
various kinds of exhaust-hoods which were placed
over them worked satisfactorily. Now the use of the
electric light renders gas foot-lights unnecessary, and
greatly decreases the amount of air required for
ventilation.

Concert-rooms may be treated on the same plan as
theatres, but owing to their being as a rule rather less
crowded, there is less difficulty in supplying the air.
Electric light being generally used, the heat from sun-
burners—which were generally used for lighting
purposes—is not now available for ventilation. Propulsion by a fan seems the best method for this class
of room, the air being passed over hot-water or steam
pipes, screened, cooled, or moistened as required, and
caused to enter the room at a height of seven or eight
feet from the floor, by a number of openings. Outlets
should be made near the ceiling, in such positions as
will allow of the air being generally diffused. It is

convenient, even if the electric light be used, to have one or two gas-lights in the roof, of the ventilating type, in case of accidents to the electric supply, as well as in the corridors. The shafts belonging to these are available as exits for the main ventilation, and the exhaust from these would be sufficient to draw a small quantity of hot air through the building for mere purposes of heating, before the room was wanted for a concert. One of the great advantages of the propulsion method for rooms of this kind is that the current of cold air from the various doors, which is so unpleasantly felt in most concert-rooms, is reduced to a minimum.

CHAPTER XIII.

THE VENTILATION OF WORKSHOPS.

THE legislature takes the sanitary condition of workshops under its supervision, and the inspectors of factories have directions with regard to gross faults in ventilation. By the Factory Acts "adequate ventilation" is prescribed. This is generally taken to signify 250 cubic feet of air per head in ordinary working hours, and 400 feet during overtime, a very insufficient amount. Three gas-burners are considered to be equivalent to one man. (This is an inadequate allowance, one gas-burner, as a rule, fouling more air than three men.)

The Factory Inspectors are not trained sanitarians, but they have the power of taking a Medical Officer of Health with them if they suspect matters to be amiss. They also have the power of ordering that a fan should be used, if there be much dust arising from the nature of the work. In wet cotton cloth factories 600 cubic feet per head are to be admitted, and the "arrangements for ventilation shall be kept in operation, subject so far as possible to the control of the persons employed

therein." Inspectors are directed to examine into the
temperature, humidity, and ventilation of such places.
The temperature must not rise above 70° Fahr., and the
difference between the wet and dry bulb thermometer
shall not be less than 2°. No specific instructions,
however, are given to the inspector as to how he is to
"examine into the state of the ventilation," no standard
of purity of air being given, or directions for measuring quantity or velocity, or for performing chemical
analysis.

One weak point about these legislative regulations
consists in the fact that none of the above provisions
are applicable to private houses, where no power is
used, and members of the family only are engaged ; and
there is great reason to believe that a large number of
such places exist, especially in the ready-made clothing
trade, where a large family, with perhaps several
apprentices, work in rooms not suited for the purpose,
amid the most grossly insanitary conditions. Ventilation in workshops is a comparatively easy matter.
There is always power available to work a fan, and
steam for heating purposes. It is only necessary to
make openings behind the steam-pipes to warm the
incoming air. In some places an exhaust is readily
made by means of a steam jet working in a properly-
made tube, though this is very noisy. An attempt to
ventilate by means of "ventilators" is always followed
by the closing of the openings in cold weather.
Printers are the most sensitive in this matter. They
have to work largely at night, with a very strong gas-
light. As a result the room becomes very hot, the men
work with very scanty clothing, and the least breath
of cold air is naturally resented. Any air which is
admitted must be warmed, and hoods arranged to take
away the fumes of the gas. The substitution of electricity for gas will be an enormous improvement, from
a sanitary point of view, in printing establishments.

CHAPTER XIV.

SUCCESSFUL VENTILATION SCHEMES.

BEFORE giving a series of examples showing the application of the above rules to actual buildings, it may be well to refer to a few buildings in which a carefully-designed ventilation system is found to give good results.

The Houses of Parliament have been referred to before. When the House of Commons is not filled with more people than it was designed to hold, the ventilation is quite satisfactory. The motive power is here a shaft in the Clock Tower heated by a furnace, and this suffices for the greater part of the year, but in the summer months it is necessary to supplement this by an exhaust fan. The air is admitted over steam radiators, and the lighting is by gas placed above a glass roof, the fumes of the burners being led into the general exhaust-shaft. Arrangements are projected for substituting electricity for gas, but as the gas fumes do not enter the rooms, it will make no difference to the ventilation.

Mention has already been made in the preceding chapter of the Opera Houses at Vienna and Frankfort-on-Main, where the air is heated by steam and propelled by a fan. In both these establishments a very perfect control is maintained, and the officer who superintends the sanitary arrangements can tell in his office the temperature of every part of the house, and lower or raise it as is required.

In warmer climates the cost of cooling the air forms a considerable part of the expense. At the Madison Square Theatre in New York the arrangements are so perfect that with a temperature outside of 80° Fahr. the air discharged after passing through the house has a temperature of only 82°. This is effected at an expenditure of six tons of ice per night.

Very little attempt has been made to mechanically

G

ventilate churches, but the Fifth Avenue Presbyterian Church in New York has been much commended by authorities for its excellent arrangements. Sir Douglas Galton speaks of it as the best-ventilated church he has ever seen. In this building the air is drawn by a fan from the top of a tower and propelled through a heating chamber containing four sets of heating coils. There exists besides a system of steam-pipes which serve to warm the floor. The latter are used for twenty-four hours before a service is held, and on the assembling of the congregation the floor-pipes are turned off and the fan set in motion.

Ventilation in most churches in England is entirely ignored. The new Church of St. Mary at Portsea, however, is an exception. A very perfect system of ventilation was projected for this church by an eminent engineer, including a fan, worked by a gas-engine, and a hot-water system for warming the air. Circumstances arose, however, in the course of the erection, which rendered it impossible to carry out this scheme, and a far inferior plan was substituted. The air is admitted through apertures in the floor, and is warmed by means of furnaces. The latter are especially large, so as to prevent the heating surfaces being over-heated, and arrangements are made for an air-supply greatly in excess of that allowed for heating purposes only. In the lofty tower a large shaft has been made, which opens at the base into the church near the roof, and numerous gas-burners are placed here to heat the shaft.

The system works, as a rule, fairly satisfactorily. Exception may be taken, however, to the plan by which the air is admitted through openings in the floor. In this case, unless the air is warmer than is necessary, unpleasant draughts are felt by those who sit near the openings : and when the whole of the furnaces are not employed, no air can be admitted from those which are not in use. Though the tower-shaft, as a rule, works efficiently, there occasionally

occur conditions of wind under which the current is
reversed, and there is actually a down draught from
the tower. In a properly-designed mechanical system
the force of the wind is combated by increasing the
velocity of the propelling or exhausting fan.   ·

The last few years have seen a great activity in the
building of large colleges, mostly for the purposes of
technical education. As these institutions frequently
include an engineering workshop, it has been easy to
arrange a system of mechanical ventilation combined
with heating by steam, and an interesting account of
the practical working of a number of schemes of this
kind is given by Mr. Robins in the work above-
mentioned.

Among schools recently built may be named a suite
of class-rooms, accommodating about 300 boys, erected
in connection with the Leeds Mechanics' Institute.[1]
Steam is used for warming, but there is no mechanical
ventilation. The outlet-shafts from the rooms are
collected and taken down to the base of a chimney-
shaft in which is placed a quantity of steam-piping.
Air is admitted to each room from the outside over
steam coils, the heat of which is controlled by the
mixing-valve arrangement described above. The upper
floor (third storey), containing a chemical laboratory,
is not included in this system, the rooms being ventilated by shafts carried well above the roof, many
of which are heated by gas. On the whole, the system
works well. The temperature of the rooms is easily
controlled, and the atmosphere is changed about three
times an hour. The chemical laboratory would no
doubt be better ventilated by a fan, but there was
great difficulty in arranging for power to actuate it.
In summer the exhaust is not in action, as the steam is
not required for heating purposes, and ventilation has
to be carried out by means of open windows. This is
the weakest point of the scheme. For this reason,

---

[1] *Vide* Fig. 13, p. 84.

even when steam is used for heating, it is better to actuate the fan by a gas-engine, or to heat the chimney by a special coke furnace. It is found also that, in rare instances, a strong wind will exercise such exhaust power on the lee side of the building as to reverse for a time the current in the exhaust-shaft.

CHAPTER XV.

EXAMPLES OF METHODS BY WHICH BAD VENTILATION MAY BE IMPROVED.

In the following pages a number of examples will be given illustrative of various kinds of buildings erected without any provision for ventilation, in which methods are proposed for the improvement of the sanitary condition. Most of these are from the actual experience of the author, and illustrate failures as well as successes. It must be clearly understood that in few of these could really perfect ventilation be secured, but so great an improvement was produced that the trouble and expense of the alterations were amply justified. It may be premised that nothing can be done without the expenditure of money, although an opinion to the contrary is very prevalent. " Ventilators " can be bought cheaply, but not ventilation.

Example 1. A church seating 400 persons, with nave and aisles, clerestory windows, some of which open ; lighted by gas-pendants ; warmed by hot-water pipes in channels on the floor ; no openings for admission or egress of air ; a western tower of moderate height.

Suggestions. (1) Make inlet apertures at the ground level on each side (two at least say 2 ft. by 18 in.) at the eastern portion of the nave, and place in front of these hot-water coils. As this will certainly overtax the boiler, arrangements should be made by which the flow to the floor-pipes can be cut off as soon as the building is warm, and the force of the fire directed to heating the " ventilating coils."

(2) Make an opening, say 3 ft. by 4 ft., from the church into the tower at the level of the roof, and enclose by means of lath and plaster a shaft running to the highest point of the tower. This may open into the belfry, or be carried to the top of the tower and capped by a cowl. Place a number of gas-jets at the base of this shaft. A door must be fitted to the opening into the church, to be kept shut when no ventilation is required.

(3) Reduce as far as is possible the consumption of gas by the following methods—

(i.) Fix a regulator on the main pipe, and remove the weights until the gas-flame is just steady.

(ii.) Substitute so far as is possible a few large burners for several small ones. This can be done with greatest advantage when the pendants or brackets are fitted with triple lights, one vertical and two horizontal. Remove entirely the two horizontal ones, and substitute for the central burner one burning half as much gas as the three together ; *e. g.* remove three burners (Bray's " Regulator," No. 3, is the most usually employed burner), each of which at one inch pressure will give the light of about seven candles, with a consumption of five feet of gas, and substitute one No. 5 "Special," which at the same pressure will burn about seven feet with the light of twenty candles, a saving of about half.

For summer ventilation, at least half the clerestory windows on each side should be made to open. The lower windows should be fitted with hoppers, so that the incoming air has an upward tendency, and advantage should be taken of any window comparatively remote from the sittings to admit as much air as possible. All windows and doors should be open for half-an-hour at least after every service ; then (in winter) closed in order to allow the building to become warm ; but the pipes which are only used for heating should be put out of the circuit before service is commenced. Air may also be conveniently admitted by making the inner porch or screen which generally protects the door open at the top, the outer door being left open, so that air is admitted about eight feet from the floor.

Rules will be found in Chapter XVI. for calculating the size of the inlet openings, the shaft, and the amount of gas required to heat it. It is most desirable that the heating apparatus in a church should be kept constantly alight. It is very difficult to properly heat a large stone building in a short time. The difference of cost between warming a building for Sunday use and keeping the fire going the whole week is only an excess of one-third in the amount of fuel consumed. This excess will to a large extent be covered by the diminished cost of keeping the organ in repair, and avoiding other damage by damp. To make a building of this kind really satisfactory, a new heating apparatus of larger capacity will be required.

Example 2. A substantially-built brick mission-room, with windows at the sides and one end; open roof; heated by a stove and lighted by gas-pendants of "star" form. This is a much more difficult room to render healthy, inasmuch as there is no tower. The stove is probably utterly insufficient to warm, if there be any adequate change of air.

For inlets, cut a large hole in the wall behind the stove, and place screens of sheet iron on each side of it, in order that the air may impinge on the heated surface. The question of having a larger stove will be considered, or additional heat may be obtained by means of gas-stoves. The "drying-room stove," by Fletcher & Co. of Warrington, is a good one for this purpose, and should be placed in front of a large opening in the wall, or the opening may be made to deliver underneath the stove. The chimney, which is small, may be run up the wall and through the roof without being unsightly. One large ventilating cowl

of the usual "exhaust" type should be placed on the ridge of the roof, in such a position that a high-power gas-burner, furnished with ventilating-tube, can be placed underneath it, the tube to run up into the cowl. Two high-power burners would be sufficient to light a room of this kind, and it would be better if both could be connected with the roof ventilator.

The windows should open with hoppers, as in the previous case, and there will be no need of any special inlet ventilators.

An iron church or school of the same type may be ventilated on similar principles. Appearance here is, as a rule, little considered, and as the building is generally a temporary one, probably the best way of heating it would be by gas-stoves of the tubular pattern, such as those made by Fletcher & Co. These should be placed against the sides of the building, near the entrance, with large apertures behind or underneath them ; the flue-pipes should be carried up inside the roof, to open at the top. At the other end of the building place an outlet cowl, connected if possible to a ventilating-pipe from one of the pendants. The inlet apertures should be closed while the building is being heated, but the convenience of gas is that it may be kept low and the building prevented from getting very cold, without the attention required by a fire. It is, however, for continuous heating, about six times more costly than coke.

Under no consideration are gas-stoves to be used to heat any inhabited building unless provided with a flue. Many stoves are sold for the purpose, as "requiring no flue." This is a fraud. So-called "condensing stoves" simply condense a little water. The foul fumes from gas cannot be condensed, and no room is healthy in which gas is burnt (for any purposes) for lengthened periods without a chimney.

As heat is nearly as keenly felt in an iron building as cold, a perforated iron pipe may be run along the ridge of the roof, and a gentle stream of water allowed to run down. The cost of the water is insignificant, and the relief great.

Example 3. An extemporized mission-room, con-structed out of two cottages. A low room, with two windows, two doors, a fireplace, lighted by two or three gas-pendants or brackets, or by paraffin lamps hung against the walls.

It is very difficult to make a room of this kind healthy or comfortable if crowded. The heat from the occupants is so great that no stove is tolerated, except in the coldest weather. Something can be done by making an aperture into the chimney at the level of the roof, or into both chimneys, if, as frequently happens from two rooms having been knocked together, there be two. One of the gas-brackets or lamps used to light the room should be fixed on the wall a little below this opening. The windows should be made to open inwards, or, if they are sashes, a hopper can easily be extemporized, by placing a piece of wood resting on the sill, hung with chains so as to incline inwards. If there be means of placing a stove (burning either coke or gas) against an outside wall, a hole should be cut in the wall behind the stove, as in the preceding example. Such rooms should never be used for more than an hour without being vacated, and all the windows opened. As a matter of fact, it is impossible to render them healthy, and from a sanitary point of view they are not to be tolerated.

Example 4. A Sunday-school class-room.   Of good height, heated by an open fire.

A good deal here depends on the kind of fireplace. With an open fire the room never becomes as offensive as when a stove or hot water is used to heat ; and an open fire is infinitely the best way of heating and ventilating a small room. But with a quick combus-tion fireplace of the ordinary type there will be, in cold weather, draughts from all the windows. With a slow combustion grate there will be insufficient ventilation. Probably the best fireplace is one of the " ventilating " or "hot-air" type, such as the Galton, Boyd, or one recently made by the Teale Fireplace Company, of moderately quick combustion. This introduces a brisk stream of warmed air, and acts as an inlet ventilator even when the fire is not lighted. The windows should be arranged to open inwards on the hopper principle, and a chimney-breast ventilator be placed at the level of the ceiling. For a room 18 ft. by 16 ft. a three-light "Stott Thorp" or "Harrison" light will be sufficient as a pendant in the centre of the room, it being understood that there is a governor on the main gas-pipe. For these lights the governor should be adjusted to give a pressure at the burner of $\frac{9}{10}$ inch (water gauge).

Example 5. A long, lofty room, with glass roof, originally built for a picture-gallery. Windows in the roof will open. Heated by hot-water pipes round three sides. Now used as a class-room for mechanical drawing. Lighted by a double row of small gas-jets, besides which there are about thirty shaded table gas-lights for the use of the students.

This room was perfectly intolerable in winter. The roof ventilators could not be opened without causing icy down draughts. The room became oppressively hot and very stuffy, and the condensed moisture fell from the glass roof in large drops on the students' drawings. The following changes were made—

1. The rows of gas-jets above were taken away and the room lighted by two high-power (100 candle) burners, furnished with reflectors and ventilating-tubes, which passed through the roof.

2. The table lights, which were of the common type, and too large (Bray's "Regulator," No. 5), were replaced by better (Bray's "Special," No. 1), which gave a sufficient light, with a consumption of three feet less gas apiece. This saved about ninety feet of gas per hour, with its attendant impurities. Two inlet ventilators were placed on the outside wall of the room, at points most remote from the students' desks.

Result. The room is by no means perfectly ventilated, but the atmosphere is quite tolerable even when it is well filled. There is no condensation of moisture, and the only gas-fumes which are not carried away are those inseparable from the table lights. Great relief has been expressed both by teachers and students. But to make it really comfortable, there is required a larger inlet for warmed air.

Example 6. A room 30 ft. by 50 ft. Used as a designing room in a school of art. Lighted by three windows at one end, and small circular lights along one side. The room is rather dark except at the end nearest the windows, but is used mostly at night. It was lighted at night by two small high-power burners at the inner end, and a number of table lights with shades over the rest of the room. The atmosphere of the room was very oppressive from the strong smell of burnt gas. It was heated by hot-water pipes. Two hit-and-miss ventilators at the window end near the floor were found (as usual) to be stopped up.

It was impossible here to dispense with the table lights, but the two high-power burners were replaced by one larger one. It was found possible to run a fourteen-inch zinc tube through the inner wall of the room close to the ceiling vertically up into the open air above the roof, one storey above, where it was capped by a cowl. Into this tube the ventilating-pipe of the larger burner was conducted, in order to cause a draught. The windows of the outer end of the room were arranged to admit air between the sashes, on the "Hinkes Bird" principle, described in an earlier chapter. The hit-and-miss ventilators on the floor were led into vertical tubes about five-and-a-half feet high.

Result. In moderate weather, especially when the windows can be kept a little open, the room is comfortable. In very cold weather there is a draught felt from the windows and inlet tubes, but not so great as might have been expected ; the descending current of cold air being partly heated by the ascending hot air from the table burners below. The room would be improved by the fixing of a hot-water coil under the windows, with a large inlet behind it. The exit-tube acts very well, and the general improvement in the room is very great.

The following example (7) illustrates well not only faults in construction, but faulty methods of improving the same, due to ignorance of the laws affecting air currents.

A lecture theatre measuring 32 ft. by 22 ft., with a high staging for students, accommodated classes of about 50. The architect had made a shaft running along the ceiling 12 in. by 8 in. in area, communicating with the room by a few small holes, and open to the air at each end. There were two outside walls, and the space below the seats, in which was a large hot-water coil, opened by a door into a passage running along one side. It is almost needless to say that there was no ventilation at all. The shaft on the roof was utterly useless. There was no inlet. Subsequently a Sherringham valve was placed near one corner of the room, but this made very little difference. A maker of ventilators was now consulted, who placed two inlet valves on the outer wall, behind the lecturer, and a third on the wall adjoining the passage. From this latter a tube about 8 in. by 7 in. was carried, with a right angle bend, through the roof into the air. Further, the shaft in the roof was connected with a long horizontal tube, which, after a bend at right angles, was taken a considerable distance vertically upwards through the roof, where it terminated in a "patent exhaust ventilator." Examination of the room after these alterations were made, showed that the ventilation was still very defective. A certain amount of air was admitted by the two valves on the outer wall of the room, but the other valve, with its tube attached, and the tube in the roof, were practically inert and useless. This is precisely what might have been expected, the length of the tube and the right-angle bends causing such friction that practically no air passed. Eventually the room was improved by further measures. The door leading to the space beneath the staging had its lower panels replaced by perforated zinc, and one of the skylights was made to open. The latter could not be kept open in winter on account of the cold down draught, but served admirably to clear the room of foul air after a lecture, before the next began ; but the result of making an opening in the door at the panel near the ground level was that a steady stream of air entered by this inlet, and becoming warmed by the hot-water coil, was not felt as a draught. The exit on this system was by one or more of the inlet-valves in the upper part of the theatre.

Example 8. The following instance of the difficulties experienced in ventilating a large reading-room and library in a public institution affords a good example of what may be done under very unpromising conditions. The room, long and lofty, formed one corner of the building, thus consisting of two parts at right angles to each other. It was lighted by a number of three-light gas-pendants, and a few table lights, and warmed (insufficiently) by hot-water coils. There were small apertures above the gas-pendants leading to tubes which passed horizontally into the air. There was a circular window near the ceiling communicating with a vestibule. When the gas was lighted and the room moderately full the atmosphere was very bad, and caused great complaint. An agent for Tobin's ventilators was then commissioned to insert some of these tubes. About twelve of these were fixed, apertures being cut through the window-frames. The windows were large casements, and caused such draughts when open that they were invariably closed. These ventilators, however, were only 9 in. by 3 in., and the tube contained two right angles, so that they were practically useless, and made no real impression on the atmosphere of the room.

The following measures were now taken :—

1. The hot-water piping in the room was greatly increased, and a special coil erected at the end of the room adjoining the vestibule. Behind this three apertures were made, one foot square each, and a wooden screen was erected in the room in front of the coil.

2. Two metal tubes, 12 in. by 6 in., were led from the ceiling level through the wall into a staircase, and vertically up to the roof, and capped by cowls. At the lower end of the tubes gas-burners were placed.

3. At opposite sides of the room, *i. e.* on the outer wall of each portion, a hopper window was constructed.

4. Gas-burners of improved construction were fitted to the pendants, and the gas supply carefully regulated.

None of these alterations except the increase of the heating surface involved much outlay, and it is instructive to note the result.

The atmosphere of the room is very greatly improved. The exit-shafts heated by gas take away a very considerable amount of foul air, but it is to be regretted that they were not thrice the size. There is a good supply of fresh warm air from the large opening at the end of the room. The principal improvement, however, is due to the large amount of hot-water piping, which so warms the room as to allow of the doors being kept open, except in very cold weather. The doors are some distance from any of the seats, and there is an inner screen. On the other hand, the hopper windows give rise to such draughts that they are hardly ever opened. In point of fact, the places were badly selected for their insertion, and there are no doubt positions in the room in which they would do good service. Were the gas replaced by electric light, there would be very little to complain of in the condition of the room, as it is not, as a rule, largely filled by occupants. As usually happens, the draught does not always set in the same direction. A strong wind against the side of the building opposite the principal inlet will drive the air out through the latter, entering by the Tobin tubes, but this does not often occur.

Example 9. A large square room capable of seating
about 100 people, and used principally for meetings
of a scientific society, about sixty in number. It was
warmed by an open fire, and lighted by a large gas
chandelier of about twenty lights. The room became
very hot and the atmosphere extremely offensive in a
short time, even when only a small number of persons
occupied it.

The first step was, as usual, to introduce some patent
ventilators. Two inlet valves were inserted, the inventor and patentee stating that they would give
sufficient air for 100 people. As these openings were
about 64 square inches each in area, the previous estimate of fresh air actually required must have been decidedly economical. A mica flap ventilator was placed
in the chimney. These measures could hardly be
said to make any marked difference in the state of the
room.

Eventually the matter was more seriously taken in
hand. The chandelier was removed and a sun-burner,
with proper ventilating-tube, substituted. An opening
2 ft. by 1 ft. was cut in an outside wall, at a point the
most remote from the fire, and a steam-coil placed in
front, and enclosed in a wooden casing lined with zinc
and felt, and open at the top. There are now two
exits, viz. the chimney, and the ventilator of the sun-
burner. The large inlet (two square feet) can be kept
open in the coldest weather, and provided there are
not more than about sixty persons in the room, the
atmosphere is fairly good.

Example 10. The waiting-room for out-patients of a
large public dispensary, accommodating about 200
persons. Twice a day this room was filled by an unwashed crowd of sick. The room was badly warmed
by two open fireplaces, one at each end. There were
two large Tobin tubes, and two shafts in the ceiling
running up to the top of the roof into a suitable cap.
There were plenty of ventilators, but no ventilation.
The Tobin tubes were invariably closed, as they caused

cold draughts, and the exhaust shafts were of no use without them.

By way of improvement the fireplaces were disused, and a large Musgrave stove placed near the centre of the room, the flue being led into one of the chimneys. An air-duct 2 ft. by 1 ft. was run under the floor from an outside wall, opening underneath the stove. The stove was purposely a very large one, in order that the fire might not require urging, and the stove never get very hot. This alteration was followed by very great improvement, large quantities of air entering by this duct, and the exhaust-tubes in the roof acted satisfactorily. Appearance in this case was not considered. Had it been necessary to avoid anything of an unsightly nature, the same effect could have been secured at greater cost, by placing hot-water coils against inlets on the outside wall. The stove, however, was by far the cheaper method.

Example 11. A dining-room in a private house, 20 ft. by 18 ft. It was lighted by a central four-light gas pendant, and there were gas-brackets on each side of the chimney-piece. There was an ordinary fireplace. The room became inconveniently warm and oppressive very soon after the gas was ignited, and there were draughts from the windows. The first method adopted was to substitute a slow combustion grate for the one in use. This did indeed mitigate the draughts, but made the room more oppressively stuffy than before. Next an inlet ventilator was inserted, but in cold weather this merely renewed the draught nuisance.

The matter was then taken in hand in more scientific fashion. The four-light pendant, consuming about twenty-four feet of gas per hour, was replaced by one Regenerative burner, which consumed six feet only, and the brackets were fitted with regulator burners, consuming four feet each instead of seven. The gas consumption was thus reduced to fourteen feet from thirty-eight, with an equal amount of light. An opening was made in the ceiling and a ventilating-pipe run into the chimney; the ventilating-tube from the central burner being connected with this. An opening was made above the inlet of the door whereby air was admitted from the hall, which was heated by a stove. This involved some structural alteration and some expense, but the saving in the consumption of gas amounted to about two shillings a week.

Example 12. A lecture hall, holding about 400 people, had and still has an evil reputation for its insufficient ventilation. It was lighted by gas chandeliers of the usual type, with nearly horizontal burners, and warmed by hot-water pipes beneath the staging.

The following measures were taken to improve matters, but with very little effect.

1. The gas-pendants were disused, and the room lighted by electricity.

2. A number of shafts (seven or eight) about 2 ft. by 1 ft. were taken from various parts of the room up to the roof, and gas-burners placed at the entrance.

3. A number of Tobin tubes were placed on the outside wall.

The result is interesting to observe as a warning. There is no arrangement for admitting warmed air from beneath, consequently the numerous exit-shafts are almost entirely useless. Some do carry off a little foul air, others act downwards and admit a small amount. The Tobin tubes are as usual always closed. The case illustrates what has been said before as to the evil of several outlets, and the uselessness of exit-shafts unless proper arrangements are made for admitting a due supply of warmed air from beneath. The most efficient way of ventilating such a room as the above would be by a small fan which should take air from the outside and propel it over the heating apparatus into the room. This could easily be actuated by an electric or hydraulic motor.

The above examples give a general idea as to the kind of measures which are likely to improve insanitary rooms, as well as those which have been found useless.

There is no royal road, and no invariably successful system. There are often three or four ways of securing ventilation, any of which may be successful. Which of them should be adopted is generally a question settled partly by pecuniary considerations, partly by. local circumstances, which vary in every case.

The greatest caution should be exercised in receiving the statements of persons, either unskilled or interested, on the merits of any ventilation arrangements in actual use. The only evidence which is of any value is the measurement by experts of the amount of air actually being passed through a building, and an analysis of the air, when the building has been occupied for at least an hour. Speaking generally, as a test for unskilled persons, the complete absence of any abnormal odour or excessive heat when the room is entered by a person fresh from the outside air is the only evidence which should be trusted. We live in an age of ventilators, it is to be hoped that this is the precursor of an age of actual ventilation. Sir Douglas Galton has well said—

"If the opinion was only equally spread through the community that bad air was detrimental to health ; if the fact of a room being close or stuffy was regarded as disgraceful ; if people refused to attend dinner-parties where the rooms were filled with bad air : the architects, the builders, and the occupiers would soon find means that every room should be pure and of a comfortable temperature."

CHAPTER XVI.

METHODS OF CALCULATION, FORMULÆ, ETC.

VENTILATION is an exact science, admitting of exact results, under varying but still quite well-known conditions. Except, however, in schemes worked out by practical engineers, where a definite steam-power is

H

used, it is very unusual to find that any sort of calculation has been made. Sizes of openings, flues, &c. are
guessed at in the roughest possible way, and builders
and designers seem to have very little idea of the real
capabilities of the apparatus they make or purchase.
Easy as calculations of this kind are to the skilled
engineer, they appal, by the very appearance of the
mathematical formulæ, those who are not accustomed
to deal with figures in this way. For a fuller treatment of this subject the reader is referred to the
engineering manuals referred to in the Appendix ; the
following examples, adapted mostly from those given
by Box (*Practical Treatise on Heat*) and Bacon (*Technical School Building:* Robins), are merely given to
show the general way in which the recognized factors
for good ventilation work are obtained.

It is most important in considering the size of flues,
inlet openings, &c., that the gratings which are usually
fixed over these for the sake of ornament and the
exclusion of birds, &c., should be so much larger than
the required area as to have an open space equal to
the required area ; *e. g.* a flue one square foot in section
is often furnished with an ornamental air-brick, whose
area of open space is barely half that of the opening.
A flue 12 in. by 12 in. generally requires a grate about
14 in. by 14 in. or larger to give the requisite space.

Example 1. Required the conditions for ventilation
of a class-room containing 40 children and a teacher.
Size 25 ft. by 24 ft. by 13 ft. ; external walls, 9-inch ;
internal, 4-inch ; 3 windows, 9 ft. by 4 ft. Air to be
admitted under the windows to the amount of 700
cubic feet per head per hour. The temperature to be
maintained at 60°, with an outside temperature of 25°,
i. e. a difference of 35°. Extraction to be by a proper
flue.

The amount of heat required is arrived at by estimating—(1) The heat lost by the walls. (2) That required
to warm the incoming air, subtracting from this the
amount of heat given off by the occupants.

1. *Loss of heat by walls.* The following table shows the results of experiments on this point—

Brickwork. B.

Thickness.	Temperature of Wall.		Units per 10°.
Inches.	Inside.	Outside.	
4½	42	36	2·312
9	45	35	1·91
14	48	34	1·59
18	49	33	1·40
27	51	32	1·10
36	53	32	0·92

Stone.

Thickness.	Temperature of Wall.		Units per 10°.
Inches.	Inside.	Outside.	
6	40	37	2·6
12	42	36	2·3
18	43	35	2·1
24	45	35	1·9
30	46	34	1·7
36	47	34	1·6

For glass the factor is ·59 per square foot per degree, or 5·9 per 10°. We may calculate then the loss of heat in the room as below—

Walls 2433 units Door 148 units
Windows 2003 ,, Floor 2755 ,, $= 7339$ units.

2. *Heat required for ventilation.* 41 persons require (700 by 41) $= 28,700$ cubic feet.

It is convenient to reduce this to lbs. for purposes of calculation, and this amount of air would weigh 2191 lbs. The heat required to raise this from 25° to 60° would be (·238 being the specific heat of air)—

2191 × ·238 (60 − 25) = 18,242 (the heat required for ventilation).

3. *Heat given off by the occupants.* This may be taken on an average as about 191 units per head. The total amount is therefore 41 × 191 = 7831. It will be seen that this figure is larger than the loss of heat from the room, so that there are really 492 units to spare for assisting the ventilation requirements. The room would therefore get hotter unless it were cooled by the incoming air. The total heat required = 7339 + 18,242 − 7831 = 17,750 units. The temperature of the incoming air would be—25 + $\dfrac{17750}{2191 \times \cdot 238}$ = 59°.

The amount of heated surface required to raise the air to the necessary height would be according to standard authorities—

α Steam-pipes at 240°   ...   ...   46 square feet.
β Hot-water pipes at 150°   ...   112   „   „
γ High-pressure hot-water pipes at
237°   61   „   „

The heating surface might be placed in a chamber below the floor, or against the walls of the room, covered in front and at the sides, open at the top, and the air admitted about 5 to 6 feet above the floor.

Extraction. The size of the extract shaft would depend on the available height, and the temperature to which it was raised.

Let us suppose it is desired to adopt a simple shaft, open to the top of the building, say 40 feet in height, not artificially heated. Taking the temperature of the shaft as 60 and the outside air at a mean of 50, we may calculate the size from the following formula—

$$D = \sqrt{\dfrac{N}{0 \cdot 85 \ \sqrt{d\,h}}}$$

Where D = Diameter of flue necessary.
    N = Number of persons.
    d = Difference between internal and external
        temperatures.

If the calculation be on the basis of allowing 1000 feet per head, the formula stands —

$$D = \sqrt{\frac{N}{0.6 \sqrt{d\,h}}}$$

Putting in the values here we have —

$$D = \sqrt{\frac{41}{0.6 \sqrt{40 \times 10}}} = \begin{array}{l}1.85 \text{ feet, or on the 700 feet} \\ \text{basis } 1.55 \text{ feet.}\end{array}$$

This will perhaps be an inconveniently large shaft. If however we can arrange to heat the exit-shaft, we shall exercise considerable economy in its size.

As an example of this and of another method of calculation, we may take a case where we have a large central exhaust heated by the chimney of the boiler passing through it, taking the room as one of 6, with a total required exhaust of 172,200 cubic feet per hour, assuming the mean temperature of the smoke-flue as 300°, and the height of the shaft as before, 40 feet ; the external diameter of the smoke-flue as 12 inches ; the mean temperature of the air entering the shaft as 54; we may calculate the mean temperature (T) of the up-cast shaft by the formula —

$$T = t + \frac{3\,s\,(t' - T')}{w}$$

Where s = Surface of heating flues (square feet). (Here 125·66 sq. feet.)
t' = Temperature of ditto.
w = Weight of air in lbs. passing up shaft per hour.
T' = Temperature of external air.

We have therefore —

$$T = 54 + \frac{3 \times 125 \cdot 66\,(300 - 54)}{6 \times 2191} = 60 \cdot 9$$

$$V = 0 \cdot 18 \sqrt{h\,d}$$

$$= 0 \cdot 18 \sqrt{40 \times 6 \cdot 9}$$

$$= 2 \cdot 99 \text{ feet per second.}$$

The area of the main shaft would be—

$0\cdot785 + \dfrac{172200}{2\cdot92 \times 60 \times 60} = 17\cdot18$, or a diameter of $4\cdot7$ ft.,

the flue occupying $\cdot785$ square feet. If we take the mean of winter temperature $39\cdot6$, a much greater advantage would be gained, while in many cases a greater height than 40 feet would be obtainable.

Example 2. A chapel with 400 occupants on floor and gallery. An allowance of 500 feet of air per head per hour, with an external temperature of 30°, and an internal temperature of 60°. Heat will be required—(1) To warm the walls. (2) To heat the air required for ventilation.

1. Area of walls about 3313 square feet, with about 41,081 units per hour. Windows, say 20 (6 ft. by 4 ft.), will lose 5904 units, together 46,985, or say roughly 47,000 units.

When the building, however, has been raised to its standard temperature, a smaller amount of heat will be required to maintain it. The walls lose only $4\cdot96$ units per foot instead of 12, and the total loss by the walls will be 16,432 units, and by the windows 5904; making for both sources 22,336 units instead of 47,000.

2. Heat required for ventilation. Allowing 500 feet per head per hour, we require 200,000 cubic feet or 15,220 lbs. of air. To raise this from 30 to 60 we require $(15,220 \times \cdot238\,(60-30)) = 708,670$ units, making with that required to keep the room warm (22,336 units) 131,006 units. From the 400 occupants 76,400 units are available, leaving 54,606 units to be supplied by the heating aparatus. This would require 356 square feet of surface heated by hot water at 150°. In this example the chapel is supposed to have galleries, and the occupants are packed rather more closely than is usual in modern churches. This state of things resembles that which prevails in a theatre, for the ventilation of which Mr. Bacon (Robins, *op. cit.*) gives the following formula—

$$V = \frac{191\, a - d}{\cdot 01817\ (T - t)}$$ Where $a =$ Number of persons present.

$d =$ Loss of heat at fixed difference of temperature per hour.

$T =$ Temperature of air at entry.

$t =$ Normal temperature of building.

$V =$ Volume of fresh air in cubic feet per hour.

Mr. Bacon points out that such buildings require a considerable amount of cooling, and recommends that the air be introduced 15° below the temperature of the room. This, however, seems very low, and would hardly be tolerated.

The case must be considered of a building of this kind only half filled. Then there would be only half as much air required, and the amount of heat needed for this purpose would be only half that calculated for the full room. On the other hand, there would be only half the heat supplied by the audience, so that practically we should find that nearly as much heat was required from the apparatus as when the room was filled. The draught-chimney in this case would require to be partially closed, or the heat by which it was actuated moderated, or in the case of a fan, the velocity diminished.

The draught-chimney, if 30 feet high, with an external summer temperature of 72°, may have the air heated 30°, or to 102°. This balances $30 \times \frac{\cdot 0707}{\cdot 0747} = 28\cdot4$ feet of external air at 72, that is to say, $1\cdot6$ feet of unbalanced pressure, which will generate theoretically a velocity of $\sqrt{1\cdot6} \times 8 = 10$ feet a second. This may be taken practically at about half, or 5 feet per second. 200,000 cubic feet of air at 72° is dilated at 102° to $200,000 \times \frac{1\cdot143}{1\cdot082} = 211,300$. Hence by the preceding formula the area of the chimney must be 58 in. by 5 in. = 11·6 square feet, or 3 ft. 5 in. square.

If it is proposed to heat this chimney by gas, 200,000 cubic feet or 14,940 lbs. of air require $(14,940 \times 30 \times ·238)$ = 101,672 units, to raise it 30°. The audience, however, will supply 76,400 (see above), leaving 25,272 units to be supplied by the gas. This, taking gas at 580 units per foot, requires 44 feet of gas per hour, costing, with gas at 3s., about 2d.

The area of the inlet openings may be calculated by a similar method to that by which the area of the outlet is obtained. We shall not be far wrong, however, if we allow an inlet space equal to the calculated outlet.

The late Dr. De Chaumont, as the result of his experiments, gave an estimate of 125 cubic feet per inch of area, as the maximum likely to be passed under ordinary circumstances. In an example worked out by Box (*Treatise on Heat*), 93 feet per inch is calculated in the case of a school ventilated by a heated air-shaft.

It will be seen that to work satisfactorily a scheme of this kind considerable skill in management will be required. The draught-chimney is working at its greatest disadvantage in the hottest weather, and will extract a much larger amount in winter. The draught in the colder weather must then be checked, preferably and most economically by restricting the amount of gas burned; the heating surface having been calculated for the coldest weather. During the warmer part of the winter season, arrangements must be made to control the heat by one of the methods described in the chapter on warming apparatus, not more than two-thirds being required as a rule. It is quite possible to make automatic self-regulating heating and ventilation machinery, and the ingenuity of American engineers has contrived very perfect ways of effecting this. The master of the house can move a dial in his hall, and the temperature of the house will be kept at the point indicated by the dial. This is done at no very great cost, and the arrangement would save a great deal of trouble.[1] But, speaking generally, a certain amount of intelligence will be required to work the apparatus which is constructed, and the provision of such intelligence is often more difficult than the planning and carrying out of the whole scheme. Still a great deal can be done if the conditions required are really understood and explained to the attendant who has to work the apparatus.

There is one point, however, connected with comfort of rooms, on which it is feared there must always be some difference of opinion, viz. the temperature. Older people like high temperatures ; younger people prefer rooms cool. It is seldom that the same temperature will suit every one : and so a school-room with an elderly teacher is often kept too hot for the efficient activity of the scholars. This difficulty must be boldly faced, and a consensus arrived at, in case of difference of opinion, as to the temperature which suits the majority.

CHAPTER XVII.

WORKS ON THE SUBJECT OF VENTILATION.

THE following works may be consulted by those who wish for further information on the subject of ventilation. The most exhaustive treatises on the subject are those by (1) Gen. Morin, *Études sur la Ventilation,* Paris. (2) Péclet, *Traité de la Chaleur,* Paris.

An interesting historical account of ventilation experiments of former years is given by (3) Tomlinson, *Warming and Ventilation,* London, 1886.

---

[1] The " Powers Thermostat," made by the Toronto Radiator Co (Toronto), seems well adapted for this purpose, and can be fitted to any form of heating apparatus.

(4) Most important practical information is contained in the *Practical Treatise on Heat,* by Thos. Box, London, 1885, which is a manual for engineers of great value ; but, from a physiological point of view, the estimate of ventilation requirements contained therein is very low, and entirely inadequate.

(5) Excellent also are the articles by Bacon in Robins' *Technical School and College Building,* London, 1887.

(6) Galton, *Healthy Dwellings,* Oxford, 1880, and articles by the same authority in the Health Exhibition Reports, may be consulted.

(7) Putnam (*The Open Fireplace in all Ages,* Boston, 1886) gives an interesting account of open fireplaces, ancient and modern, and American methods of heating by hot air, with reference to ventilation.

(8) *The Principles of Warming and Ventilation,* by Dr. J. S. Billings, London, 1884, and New York, 1893, is one of the most valuable works on the subject ; but the examples and general methods recommended are adapted to the climate of the northern States of America rather than that of the British Isles.

(9) *Carnelly.* Report on the cost and efficiency of the heating and ventilation of schools. (323 schools examined.) Winter, Duncan & Co., Dundee. 1889.

(10) *Ritchie.* Treatise on Ventilation. 1862.

(11) *Dictionary of Engineering* (Spon.). Art. Ventilation.

(12) *Shaw.* Warming and Ventilation. A very able article, from the engineer's point of view, in *Treatise on Hygiene and Public Health,* by Stevenson & Murphy. 1892. Vol. I.

(13) *Drysdale and Hayward.* Health and Comfort in House-building. 1872.

(14) *Phipson. Proceedings of Institute of Civil Engineers.* Vol. IV., 1879, p. 124. On the heating and ventilation of Glasgow University.

(15) *Constantine. Warming and Ventilation.* This is practically a trade advertisement of a particular kind

of heating apparatus, but incidentally some interesting information is given respecting the ventilation of public buildings in Manchester.

(16) *Engineering Record* (American). 1892. For schemes of automatic temperature regulation in connection with mechanical ventilation.

APPENDIX.

Conspectus of the state of the air in 85 schools mechanically and "naturally" ventilated, by the late Prof. Carnelly (*Journal of Pathology*, Nov. 1893).

TABLE I.

	Rooms examined.	Percentage of windows open.	CO_2 in vols. per 10,000.	Micro-organisms per litre.
Natural ventilation:				
Country	45	24	16·1	76
Suburbs and country towns	46	32	16·7	103
Town (Aberdeen) ...	42	28	18·8	136
Town (Dundee) ...	39	22	18·6	152
Mechanical ventilation:				
Town (Aberdeen) ...	12	0	12·3	20
Town (Dundee) ...	25	3	12·3	17

TABLE II.

Results classified according to method of heating.

	Rooms examined.	Windows open, per cent.	Excess of organic matter.	CO_2 in vols. per 10,000.	Micro-organisms per litre.
Mechanical ventilation (hot pipes)	32	1·5	1·1	12·3	18·5
Hot pipes and natural ventilation	43	58	6	16	78
Open fires and natural ventilation	84	58	7·5	19	158

TABLE III.

Relation of the smell of the air to the results of the analyses.

	Carbonic acid.	Excess of organic matter.	Micro-organisms per litre.
Very unpleasant ...	22·4	1·97	150
Moderately unpleasant	16·7	1·34	96
Fairly fresh	13·2	·88	83

Richard Clay & Sons, Limited, London & Bungay.

PUBLICATIONS

OF THE

Society for Promoting Christian Knowledge.

MANUALS OF HEALTH.

Fcap. 8vo, 128 pages, Limp Cloth, price 1s. each.

HEALTH AND OCCUPATION. By Sir B. W. RICHARDSON, F.R.S., M.D.

HABITATION IN RELATION TO HEALTH (The). By F. S. B. CHAUMONT, M.D., F.R.S.

ON PERSONAL CARE OF HEALTH. By the late E. A. PARKES, M.D., F.R.S.

WATER, AIR, AND DISINFECTANTS. By W. NOEL HARTLEY, Esq.

HEROES OF SCIENCE.

Crown 8vo, Cloth boards, 4s. each.

ASTRONOMERS. By E. J. C. MORTON, B.A.

BOTANISTS, ZOOLOGISTS, AND GEOLOGISTS. By Professor P. MARTIN DUNCAN, F.R.S., &c.

CHEMISTS. By M. M. PATTISON MUIR, Esq., F.R.S.E.

MECHANICIANS. By T. C. LEWIS, M.A.

PHYSICISTS. By W. GARNETT, Esq. M.A.

SPECIFIC SUBJECTS.

Fcap. 8vo, 64 pages, Limp Cloth, price 4d. each.

ALGEBRA. By W. H. H. HUDSON, M.A.

Answers to the Examples given in the above, *Limp cloth,* 6d.

EUCLID. Books 1 and 2. Edited by W. H. H. HUDSON, M.A.

ELEMENTARY MECHANICS. By W. GARNETT, M.A., Demonstrator of Experimental Physics at Cambridge.

PHYSICAL GEOGRAPHY. By the Rev. T. G. BONNEY, F.G.S.

MANUALS OF ELEMENTARY SCIENCE.

Fcap. 8vo, 128 pages, with Illustrations, Limp Cloth, 1s. each.

PHYSIOLOGY. By F. LE GROS CLARKE, F.R.S., St. Thomas's Hospital.

GEOLOGY. By the Rev. T. G. BONNEY, M.A., F.G.S., Fellow and late Tutor of St. John's College, Cambridge.

CHEMISTRY. By ALBERT J. BERNAYS.

ASTRONOMY. By W. H. M. CHRISTIE, M.A., the Royal Observatory, Greenwich.

BOTANY. By the late Professor ROBERT BENTLEY.

ZOOLOGY. By ALFRED NEWTON, M.A., F.R.S., Professor of Zoology in the University of Cambridge.

MATTER AND MOTION. By the late J. CLERK MAXWELL, M.A., Trinity College, Cambridge.

SPECTROSCOPE, THE, AND ITS WORK. By the late RICHARD A. PROCTOR.

CRYSTALLOGRAPHY. By HENRY PALIN GURNEY, M.A., Clare College, Cambridge.

ELECTRICITY. By the late Prof. FLEEMING JENKIN.

THE ROMANCE OF SCIENCE.

Post 8vo.   With numerous Illustrations.   Cloth boards.

COAL, AND WHAT WE GET FROM IT.
By Professor R. MELDOLA, F.R.S., F.I.C.   2s. 6d.

COLOUR MEASUREMENT AND MIXTURE.
By Captain W. de W. ABNEY, C.B., R.E., F.R.S.   2s. 6d.

DISEASES OF PLANTS.
By Professor MARSHALL WARD.   2s. 6d.

OUR SECRET FRIENDS AND FOES.
By PERCY FARADAY FRANKLAND, PH.D., F.R.S.   2s. 6d.

SOAP-BUBBLES, AND THE FORCES WHICH MOULD THEM.
By C. V. BOYS, A.R.S.M., F.R.S.   2s. 6d.

SPINNING TOPS.
By Professor J. PERRY, M.E., F.R.S.   2s. 6d.

TIME AND TIDE: a Romance of the Moon.
By Sir ROBERT S. BALL.   2s. 6d.

THE MAKING OF FLOWERS.
By the Rev. Professor G. HENSLOW, M.A., F.L.S., F.G.S.   2s. 6d.

THE STORY OF A TINDER-BOX.
By the late C. MEYMOTT TIDY, M.B., M.S.   2s.

THE BIRTH AND GROWTH OF WORLDS.
A Lecture by Professor A. H. GREEN, M.A., F.R.S.   1s.

NATURAL HISTORY RAMBLES.

*Fcap. 8vo. With numerous Woodcuts. Cloth
boards, 2s. 6d. each.*

IN SEARCH OF MINERALS.
By the late D. T. ANSTED, M.A., F.R.S.

LAKES AND RIVERS.
By C. O. GROOM NAPIER, F.G.S.

LANE AND FIELD.
By the late Rev. J. G. WOOD, M.A.

MOUNTAIN AND MOOR.
By J. E. TAYLOR, F.L.S., F.G.S., Editor of "Science-Gossip."

PONDS AND DITCHES.
By M. C. COOKE, M.A., LL.D.

THE SEA-SHORE.
By Professor P. MARTIN DUNCAN, M.B. (London), F.R.S.

THE WOODLANDS.
By M. C. COOKE, M.A., LL.D., Author of "Freaks and Marvels
of Plant Life," &c.

UNDERGROUND.
By J. E. TAYLOR, F.L.S., F.G.S.

LONDON: NORTHUMBERLAND AVENUE, W.C.;
43, QUEEN VICTORIA STREET, E.C.
BRIGHTON: 135, NORTH STREET.

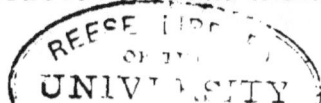

www.ingramcontent.com/pod-product-compliance
Lightning Source LLC
Chambersburg PA
CBHW021936190326
41519CB00009B/1030